Prachai Norajitra

Divertor Development for a Future Fusion Power Plant

Divertor Development for a Future Fusion Power Plant

by
Prachai Norajitra

Dissertation, Karlsruher Institut für Technologie
Fakultät für Maschinenbau
Tag der mündlichen Prüfung: 19. Juli 2011

Impressum

Karlsruher Institut für Technologie (KIT)
KIT Scientific Publishing
Straße am Forum 2
D-76131 Karlsruhe
www.ksp.kit.edu

KIT – Universität des Landes Baden-Württemberg und nationales
Forschungszentrum in der Helmholtz-Gemeinschaft

KIT Scientific Publishing 2011
Print on Demand

ISSN 2192-9963
ISBN 978-3-86644-738-7

Divertor Development for a Future Fusion Power Plant

Zur Erlangung des akademischen Grades
Doktor der Ingenieurwissenschaften
der Fakultät für Maschinenbau
Karlsruher Institut für Technologie (KIT)

genehmigte
Dissertation

von

Dipl.-Ing. Prachai Norajitra
aus Karlsdorf-Neuthard

Tag der mündlichen Prüfung: 19.07.2011
Hauptreferent: Prof. Dr. Oliver Kraft
Korreferent: Prof. Dr. Jürgen Haußelt
Korreferent: Prof. Dr. Robert Stieglitz

Abstract

Nuclear fusion is considered as a future source of sustainable energy supply. In the first chapter, the physical principle of magnetic plasma confinement, and the function of a tokamak are described. Since the discovery of the H-mode in ASDEX experiment "Divertor I" in 1982, the divertor has been an integral part of all modern tokamaks and stellarators, not least the ITER machine.

The goal of this work is to develop a feasible divertor design for a fusion power plant to be built after ITER. This task is particularly challenging because a fusion power plant formulates much greater demands on the structural material and the design than ITER in terms of neutron wall load and radiation.

In chapter 2, several divertor concepts proposed in the literature e.g. the Power Plant Conceptual Study (PPCS) using different coolants are reviewed and analyzed with respect to their performance. As a result helium cooled divertor concept exhibited the best potential to come up to the highest safety requirements and therefore has been chosen for the design process. From the third chapter the necessary steps towards this goal are described. First, the boundary conditions for the arrangement of a divertor with respect to the fusion plasma are discussed, as this determines the main thermal and neutronic load parameters. Based on the loads material selection criteria are inherently formulated.

In the next step, the reference design is defined (chapter 3.6) in accordance with the established functional design specifications. The developed concept is of modular nature and consists of cooling fingers of tungsten using an impingement cooling in order to achieve a heat dissipation of 10 MW/m^2. In the next step, the design was subjected to the thermal-hydraulic and thermo-mechanical calculations (chapter 3.8) in order to analyze and improve the performance and the manufacturing technologies. Based on these results, a prototype was produced and experimentally tested on their cooling capacity, their thermo-cyclic loading behavior and manufacturing processes (chapter 3.9). Prototypical power densities were used, which were generated by an electron beam. Chapter 4 discusses the first steps of development of the manufacturing processes for tungsten divertor components with respect to achieving micro-crack-free surface quality and the mass production.

The developed divertor concept has demonstrated its principal functionality and hence the used design process and tools can be conceived as verified and validated. Nevertheless, a large effort still has to be spent to improve the design in terms of robustness against thermomechanical load cycling to enhance its lifetime.

Divertorentwicklung für einen zukünftigen Fusionsleistungsreaktor

Kurzfassung

Die Kernfusion wird als zukünftige Quelle für eine nachhaltige Energieversorgung angesehen. Im ersten Kapitel wird einleitend das physikalische Prinzip des magnetischen Plasmaeinschlusses sowie die Funktion eines Tokamaks erläutert. Seit der Entdeckung der H-Mode im ASDEX Experiment „Divertor I" im Jahre 1982 gehört der Divertor wegen seiner guten Plasmareinigungseigenschaft zum festen Bestandteil aller heutigen Tokamaks und Stellaratoren, sowie nicht zuletzt der ITER-Maschine.

Ziel dieser Arbeit ist es, ein praktikables Divertordesign für ein Fusionskraftwerk, das nach ITER gebaut werden soll, zu entwickeln. Diese Aufgabe ist eine besondere Herausforderung, weil ein Fusionsleistungsreaktor viel höhere Anforderungen an die Strukturmaterialien und das Design in Bezug auf die Neutronenwandbelastung und Strahlung als ITER stellt.

Im Kapitel 2 wird eine Vielzahl der in der Literatur, wie z.B. der Power-Plant Conceptual Study (PPCS), vorgeschlagenen Divertorkonzepte mit unterschiedlichen Kühlmitteln in Bezug auf ihre Leistung begutachtet und analysiert. Als Ergebnis zeigte das Helium gekühlte Divertorkonzept das beste Potential hinsichtlich der höchsten Sicherheitsansprüche und wurde daher für den Design-Prozess ausgewählt.

Ab dem dritten Kapitel werden die notwendigen Schritte zur Erreichung dieses Ziels werden beschrieben. Zunächst werden die Randbedingungen für die Anbringung eines Divertors in Bezug auf das Fusionsplasma diskutiert, da dies die wichtigsten thermischen und neutronischen Belastungsparameter bestimmt. Basierend auf den Belastungen werden die Materialauswahlkriterien grundsätzlich formuliert.

Im nächsten Schritt wird das Referenz-Design (Kapitel 3.6) im Einklang mit den erstellten funktionalen Design-Spezifikationen definiert. Das entwickelte Konzept ist von modularer Art und besteht aus Kühlfingern aus Wolfram, die eine Prallkühlung verwenden, um eine Wärmeabfuhr von 10 MW/m^2 zu erreichen. Im nächsten Schritt

(Kapitel 3.8) wurde das Design den thermisch-hydraulische und thermo-mechanischen Berechnungen unterzogen, um es in Bezug auf Leistung und Fertigungstechnologien zu analysieren und zu verbessern. Basierend auf diesen Ergebnissen wurde ein Prototyp hergestellt und experimentell auf ihre Kühlleistung, ihr thermo-zyklisches Belastungsverhalten und die Fertigungsverfahren getestet (Kapitel 3.9). Prototypische Leistungsdichten wurden verwendet, die durch einen Elektronenstrahl erzeugt wurden. Im Kapitel 4 werden die ersten Schritte der Entwicklung des Herstellungsprozesses für Wolfram Divertor Komponenten in Bezug auf die Erreichung der mikrorissfreie Oberflächenqualität und die Serienproduktion beschrieben.

Das entwickelte Divertorkonzept hat seine wichtigste Funktionalität demonstriert und die verwendeten Design-Prozesse und Tools können als verifiziert und validiert erachtet werden. Dennoch, eine große Anstrengung muss noch ausgegeben werden, um das Design in Bezug auf die Robustheit gegen thermomechanischen Lastwechsel zu verbessern, um seine Lebensdauer zu erhöhen.

Table of contents

1   Introduction and Motivation

1.1   Why Nuclear Fusion?

Over the past few decades, global climate changes, such as a global ground-level temperature increase between 1906–2005 by about 0.7 °C [1-1] was observed (Figure 1-1). As the main cause of this a huge greenhouse gas[1] emissions by unrestrained energy use, especially in industrialized countries is suspected. This development is further enhanced by a world population growth over the long term with a forecast to 2050 by about 1 %/year [1-2]. One of the main objectives of the current energy policy is therefore the reduction of greenhouse gases especially CO_2 emissions. At the same time, a long-term reliable and affordable energy supply must be ensured, taking into account the finite nature of non-renewable resources[2] (Figure 1-2). Choosing the right future energy sources and strategies for sustainable energy industry is a difficult task that depends on political decisions and social acceptance. So, for example, in 1991, the German major project SNR-300 fast breeder in Kalkar was abandoned because of safety concerns and reasons of economy [1-3]. This political decision was made just at the time when the two reactor accidents in Harrisburg 1979 and Chernobyl in 1986 were still fresh in memory. After the first energy crisis in 1973/1974, when the OPEC countries ceased oil production, the importance of energy independence grabbed attention of politicians and sections of the society and decisively influenced the energy policy orientation. According to statistics from the Federal Ministry of Economics and Technology (BMWi) [1-4], the primary energy in total electricity production in Germany in 2008 was as follows: 48.2 % coal, 29.5 % nuclear energy (produced by nuclear fission of uranium-235), 15.4 % oil / gas, 4.3 % water and wind power, 2.6 % other. The use of nuclear energy thereunder allows low CO_2 emissions. This environmental benefit is met at the unsolved problem of disposal of radioactive waste, assuming no public acceptance of artificial nuclear waste transmutation as a technical waste management option.

---

[1] Greenhouse gases include: CO, CO_2, CH_4, CFCs, O_3, N_2O, as measured in CO_2 equivalents.
[2] Known and probable reserves: oil about 200 years, coal 300 years, uranium, about 200 years.

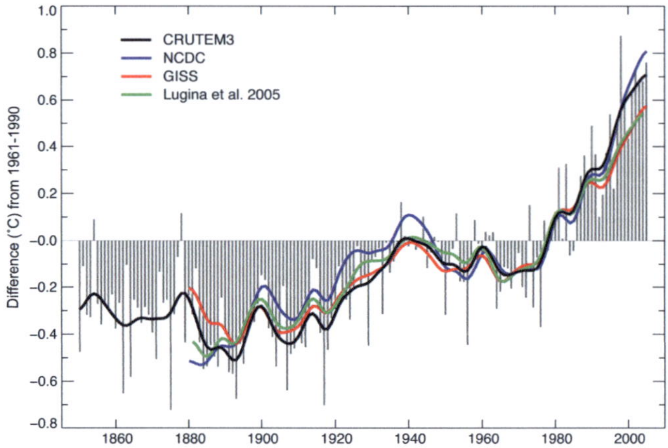

Figure 1-1: Annual anomalies of global land-surface air temperature (°C), 1850 to 2005, relative to the 1961 to 1990 mean according to different variations [1-1].

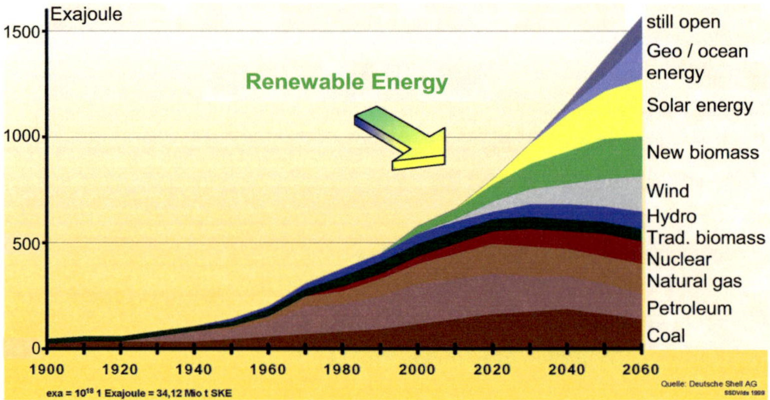

Figure 1-2: World energy consumption by 2060. Scenario: sustainable growth (1 Exajoule = 10^{18} J = 34.12 million tons SKE (coal equivalent)).

Another type of nuclear power generation is nuclear fusion, in which two light nuclei combine to form a single heavier nucleus (Figure 1-3) with the release of a large amount of energy. It is a promising option for future energy supply, which is characterized by almost inexhaustible fuel supplies, favorable safety characteristics, and environmental compatibility. In nuclear fusion, large amounts of energy from hydrogen[3] and lithium alone can be produced without emitting carbon dioxide. Furthermore, only small radioactive waste is generated with short half-lives, requiring only about 100 years storage. These advantages are great hopes on nuclear fusion as one of the most promising energy supplier, said the German Federal Government in 2010 for research projects particularly modern forms of energy production supports [1-5].

1.2 Principles of Nuclear Fusion

The energy generation from nuclear reactions – both in the fusion and in the nuclear fission – is based on the physical principle that a difference of the total mass of the particles before the reaction and the total mass of the particles after the reaction occurs. This mass loss – called mass defect – corresponds to the binding energy of the nucleus via Einstein's mass-energy relation:

$$E \quad = \quad m{\cdot}c^2 \tag{1-1},$$

with m: mass [kg], E: energy [J], c: speed of light = 299792458 m/s. This is the energy that is released during the assembly of a nucleus from its individual nucleons[4]. In other words, this is the energy that must be spent to dismantle the nucleus into its individual nucleons.

The average binding energy per nucleon in MeV (1 MeV = $1.602{\cdot}10^{-13}$ J) is shown in Figure 1-3 as a function of mass number [1-6]. Basically there are two possibilities for the use of nuclear energy. The range of very light nuclei – possible for nuclear fusion – is to the left of the absolute maximum of curve, which is at a mass number of about 60. The important fusion reactions, as the following discussed deuterium (D)-tritium (T) reaction, take advantage of the strong local maximum in the ^{4}He isotope. On the right branch of the curve maximum, heavy nuclei such as ^{235}U are split in medium heavy nuclei, such as ^{94}Kr and ^{139}Ba shown for example in the figure as possible reaction products.

---

[3] One gram of hydrogen results in the fusion, an energy equivalent of about 10,000 liters of fuel oil or 11 tons of coal.

[4] The nucleus consists of protons and neutrons, which are sometimes collectively called nucleons. The rules for spelling the elements are as follows: ^A_Z Symbol, with A: the mass number, Z: the atomic number or proton number; where A = N + Z, with N: the neutron number of a nucleus.

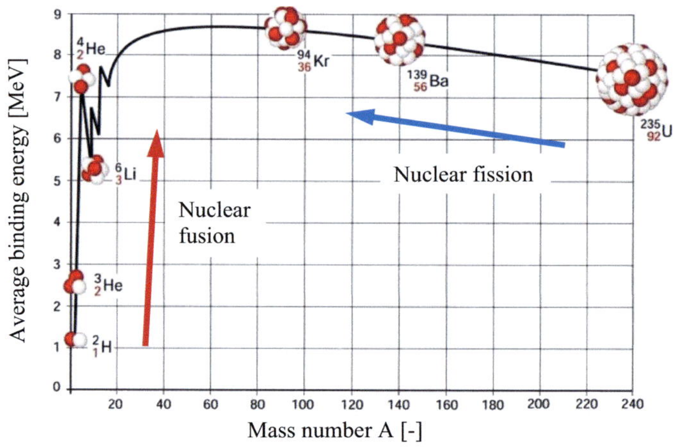

Figure 1-3: Average binding energy per nucleon as a function of mass number [1-6].

A physical fact that approaching positively charged atomic nuclei exert Coulomb repelling forces [1-6] makes the fusion process seemingly impossible. The Coulomb force increases with decreasing distance r between the nuclei according to the law 1/r. Only below a distance r of about 10^{-14} m the Coulomb force is overcompensated by the much more attractive nuclear force, so that nuclear reactions are possible. To overcome the Coulomb Walls the particles would need, according to classical mechanics, a minimum kinetic energy[5] of e.g. 300 keV[6] for a deuterium-deuterium (D-D) reaction. This energy corresponds to an unrealistically high temperature of about 3 billion degrees Kelvin. For comparison, in the interior of the sun there is a much lower temperature of around 15 million degrees Kelvin [1-7] with an average thermal energy kT of the particles of about 1.3 keV. According to quantum mechanics, however, even at lower particle energy there is a certain probability of tunneling through such a barrier. To overcome the Coulomb repulsion in the DT fusion an average temperature of about 100 million Kelvin is necessary. This corresponds to an average thermal energy of the deuterons and tritons of about 10 keV

---

[5] Approximate calculation of the Coulomb-wall height [1-8]: V_c [MeV] $\approx Z_1.Z_2/A^{1/3}$, with Z_1 and Z_2: the atomic numbers of projectile, A: mass number of target.
[6] 1 keV corresponds to about 10 million degrees Kelvin

for the tunneling through. At these temperatures, which are far above the ionization energy of hydrogen atom of 13.6 eV, the fusion reactants are in a plasma state[7].

Among many possible fusion reactions (Figure 1-4), the fusion of the hydrogen isotopes deuterium and tritium is favored:

$$\,^{2}_{1}D \;+\; \,^{3}_{1}T \;\;\rightarrow\;\; \,^{4}_{2}He \;+\; \,^{1}_{0}n \;+\; 17.6\,\text{MeV} \tag{1-2},$$

where n is a neutron. This is due to its higher efficiency and better feasibility of the plasma temperature of 100 million degrees Kelvin. The result is energy of the reaction products of a total of 17.6 MeV released, which is composed of the kinetic energy of the helium core of 3.5 MeV and the neutron of 14.1 MeV.

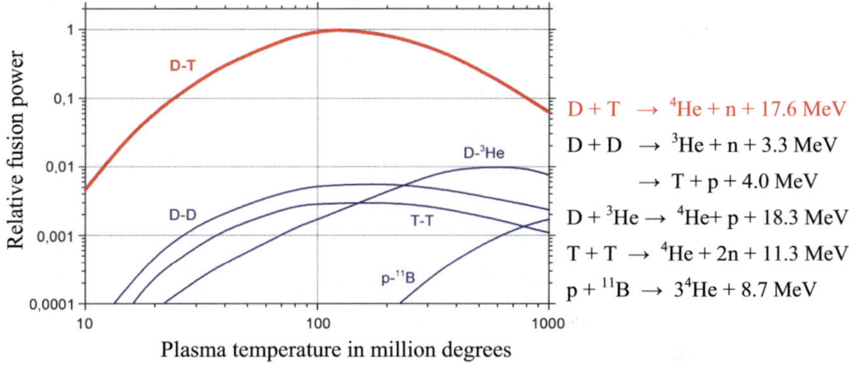

Figure 1-4: Major fusion reactions [1-6]. Optimal is the D-T reaction at about 100 million degrees plasma temperature. Abbreviation: D – Deuterium, T – Tritium, He – Helium, B – Bor, p – Proton.

The resulting fusion energy is about 10^{6} times greater than that of the chemical processes. About 30 million kWh of electrical energy can be obtained from 1 kg DT. This corresponds to the energy equivalent of about 3.5 million kg of coal or about 2.5 million liters of fuel oil [1-9]. This comparison shows that the fusion is a hope for the future energy supply.

---

[7] Hydrogen is between 0–14 K in the solid, 14–20 K in the liquid and 20–10000 K in the gaseous molecular state. Between 10,000 and 20,000 K, there is a gradual separation of the electrons from the nuclei, i.e. the atoms are ionized. Above 20000 K the hydrogen enters the plasma state, in which the hydrogen gas has become a mixture of two gases, the ion gas and the electron gas [1-7].

1.3   Magnetic Plasma Confinement and the D-T Plasma Ignition Conditions

The goal of fusion research is to produce a self-sustaining fusion plasma after a single injection of ignition – without another external power supply. In the ignited plasma state, only the consumed fuels, D and T, must be replenished. The reaction volume must be thermally sufficiently well insulated to the outside, so that this state is possible at a constant required temperature of 100 million Kelvin. Otherwise, the heat power produced in the plasma is not enough to cover the heat loss by heat conduction and heat radiation, and the plasma goes out. Therefore, any contact of the fusion plasma with the container walls must be avoided. Technically, nuclear fusion research is carried out today in two main directions based on the inertial and the magnetic confinement.

1.3.1   The Magnetic Confinement

The magnetic confinement is based on the physical property of moving ions and electrons in a uniform magnetic field on spiral orbits along magnetic field lines in the sense of a left- or right-handed screw. The circular path radius – called gyration radius – can be determined from the equilibrium between the centrifugal force ($m \cdot v^2/r_g$) and the Lorentz force ($q \cdot v \cdot B$) acting in a magnetic field on charge carriers:

$$m \cdot v^2/r_g \qquad = \qquad q \cdot v \cdot B \qquad\qquad (1\text{-}3),$$

yielding
$$r_g \qquad = \qquad m \cdot v/(q \cdot B) \qquad\qquad (1\text{-}4),$$

with r_g: radius of gyration [m], m: mass [kg], v: velocity perpendicular to B [m/s], q: charge of the carrier $= \pm\, 1.602 \cdot 10^{-19}$ [As], B: magnetic flux density [Tesla or Vs/m^2]. It is for example for deuterium with $v \approx 106$ m/s (T = 108 K), B $\approx$ 4 Tesla at about $5 \cdot 10^{-3}$ m. Assuming a mean thermal (kinetic) energy of the positive and negative charge carriers, ions (i) and electrons (e), in a thermal plasma of

$$m_e \cdot v_e^{\,2} / 2 \qquad = \qquad m_i \cdot v_i^{\,2} / 2 \qquad\qquad (1\text{-}5),$$

equation 1-4 yields the ratio of the radii of gyration of ion and electron of $\sqrt{\dfrac{m_i}{m_e}}$ of about 67.

By magnetic confinement, the particles are strongly restricted in their mobility across the magnetic field and essentially follow the magnetic field lines. Next is a requirement for them also not to leave the reaction volume along the magnetic field.

This is only possible if a toroidal plasma vessel is used with applied annular closed magnetic fields. Here, the ions and electrons can continue to move in a spiral-shaped path along the magnetic field lines. However, the curvature of the magnetic field lines leads to an inhomogeneity of the magnetic field, whose strength in the plasma cross section on the inside is greater than on the outside. As a result, the Lorentz force acting perpendicular to the magnetic field lines varies and thus additional vertical drift motion of the particle arise. This so-called gradient drift can be compensated by means of the screwing of the magnetic field lines around the plasma axis. That can be realized by superimposing the toroidal main magnetic field with the poloidal magnetic self-field of the flowing plasma current.

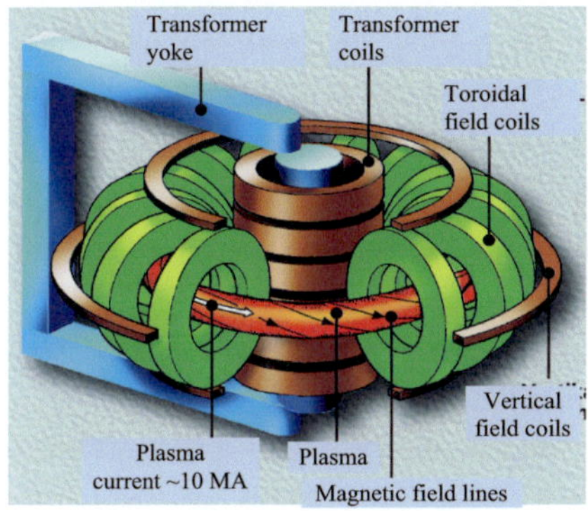

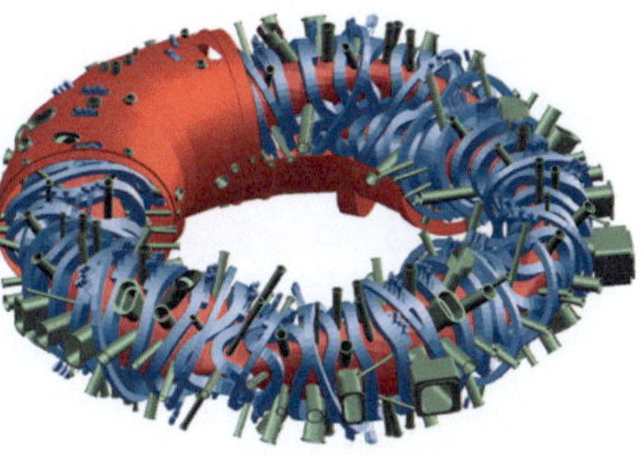

Tokamak: The plasma current generates part of the magnetic field. It has a simple geometry, but current-driven instabilities.

Stellarator: Magnetic fields are generated exclusively by external coils. There are no current driven instabilities. It has the intrinsic property of steady state operation, but complex geometry.

Figure 1-5: Principle of a tokamak [1-6] and a stellarator [1-16].

Such a toroidal plasma machine in which the screwing of the main magnetic field is achieved by the magnetic self-field of a flowing current in the plasma is called tokamak (Russian for "toroidal chamber in magnetic coils") (Figure 1-5, left). The required plasma current is generated by a transformer and provides simultaneously for the initial heating (resistive heating) of the plasma.

An alternative to the tokamak is the stellarator (Figure 1-5, right). In a stellarator the screwing of the magnetic field is produced by external coils, i.e. without the plasma current. Unlike the tokamak, its magnetic field is no longer axisymmetric and the cross section of its magnetic surface is not circular. The magnetic field generating coils and the plasma have a more complicated form.

1.3.2 The D-T Plasma Ignition Conditions

In a stationary working fusion reactor, the plasma temperature must be kept constant in time. This means that the energy loss which flows continuously from the plasma due to heat conduction (chapter 1.4.1) and bremsstrahlung (chapter 1.4.2) must be replaced by an equally large energy flow. In the DT reaction (eq. 1-2) α-particle with an energy of 3.5 MeV and neutron with an energy of 14.1 MeV are simultaneously generated. While the electrically neutral neutrons freely fly through the plasma and are slowed down in solid matter outside of the plasma, the positively charged α-particles are trapped in the magnetic field and thus in plasma. They thus serve as an internal heat source for the self-heating of the fusion plasma to compensate for energy losses.

For the ignition and maintenance of a nuclear fusion three parameters are crucial: the plasma pressure p, temperature T and the energy confinement time τ_E. The confinement time is a measure of the quality of the thermal insulation of the plasma. It is representative of how long the state of the plasma can be maintained without energy supply. To achieve a high reaction rate R_{12} for tunneling through, sufficiently high particle density (i.e. plasma pressure) and temperature of the plasma are required [1-6]:

$$R_{12}\,[1/s] \quad = \quad n_1 \cdot n_2 \cdot \sigma_F \cdot v \qquad (1\text{-}6),$$

with n_1, n_2: nuclei of varieties 1 and 2 (here: D and T) per unit volume [m^{-3}], σ_F: Fusion cross-section[8] [m^2], v: relative velocity of the nuclei 1 and 2 [m/s].

At a temperature for tunneling through the Coulomb barrier of about 10 keV (about 100 million K) (see chapter 1.3) is the Lawson criterion for $n_e \cdot \tau_E$ product [1-10]:

$$n_e \cdot \tau_E \quad \geq \quad 10^{20}\,[\text{s} \cdot \text{m}^{-3}] \qquad (1\text{-}7),$$

with n_e: electron density [m^{-3}] = sum of deuterons n_D and tritons n_T densities.

---

[8] The cross section [1 m^2 = 10^{28} barn] is a measure of the probability of the occurrence of a fusion process and depends strongly on the relative velocity of the reactants.

It says that at that particular temperature, the product of density and energy confinement time must have at least the value of 10^{20} [s·m^{-3}] in order to enable a self-burning plasma. Below $kT \approx 4$ keV, the ignition of a D-T plasma is physically impossible, because the plasma emits additional power by heat conduction.

Multiplying the sizes of the Lawson criterion with temperature, attributed it to a greater weighting, we obtain the so-called triple product or fusion product. In practice, the so derived ignition condition is used:

$$n_e \cdot \tau_E \cdot T \;\geq\; 6 \cdot 10^{21} \;[\text{keV} \cdot \text{s} \cdot \text{m}^{-3}] \tag{1-8},$$

with n_e: electron density [m^{-3}], τ_E: Energy confinement time [s] and T in [keV]. Following practical values are given in [1-6]: T > 100 million K (or > 10 keV), n_e ~10^{20} particles/m^3, τ_E ~6.0 s.

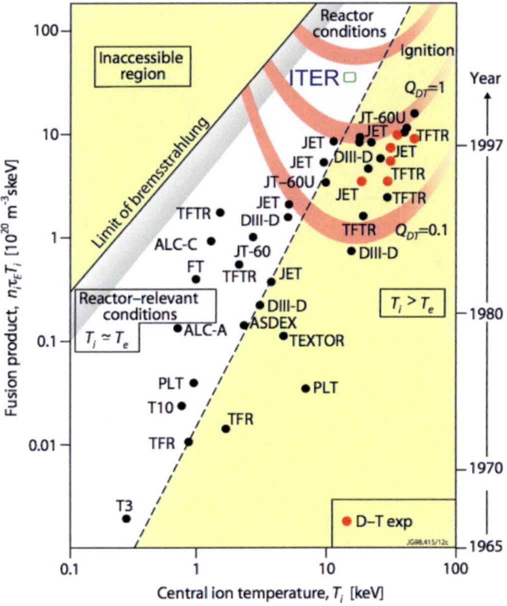

Figure 1-6: Fusion product development [1-12] [1-13].

Figure 1-6 shows the evolution of the fusion product over the time since the beginning of fusion research. It is easy to see that the goal of a burning plasma is almost reached. So far, there are some experimental facilities around the world managed to reach the break-even point[9] for an extremely short time. At this point, the ratio of energy gain and loss is equal to one. The projected International Thermonuclear Experimental Reactor (ITER) [1-11] is expected to reach the ignition and the state of a self-burning plasma. Then, a power reactor only works effectively if the plasma permanently burns far beyond the break-even point.

1.4 Energy Loss of the Plasma

1.4.1 Energy Loss by Heat Conduction and Diffusion

With the ring-shaped toroidal plasma confinement, the thermal insulation of a plasma seems to be ideal. A migration of the thermal plasma particle energy perpendicular to the magnetic field lines seems at first impossible. In fact, there is an inevitable physical mechanism that enables the transport of energy (heat conduction) and particles (diffusion) perpendicular to the magnetic field lines. This mechanism arises from the mutual Coulomb repulsion during the flyby of two plasma ions spirally moving toward each other. This will enable drifting of the path guiding centers of the particles and at the same time a transfer of the kinetic energy from the faster to the slower particles. As a result, the radius of gyration changes. By the Coulomb collisions the energy and its carrier can now move across the magnetic field from the inside outwards. As with any heat transport, a temperature gradient arises from the plasma center towards the plasma edge. This means that the temperature is highest in the plasma center and decreases radially towards the outside. The shape of the temperature profile depends on the spatial distribution of heating energy source, the local density and thermal conductivity. As equivalent to the thermal conductivity of the plasma, energy confinement time τ_E is used. It is a measure of the quality of the insulation of the plasma in a magnetic field, indicating that a large confinement time means good insulation and vice versa. In a plasma that emits energy only by conduction, the mean temperature (averaged over the plasma cross-section) is in equilibrium so that the following relationship holds [1-7]:

Fed power density into the plasma = Thermal plasma energy density / τ_E (1-9),

---

[9] The point at which the heating energy put into the plasma is equaled by the energy produced by the fusion of atomic nuclei.

with thermal plasma energy density = $3 \cdot n \cdot kT$, taking into account that two gases, ions and electrons, of density n, and temperature T are in the plasma.

1.4.2 Energy Loss by Bremsstrahlung and Synchrotron Radiation

In addition to the outflow of thermal plasma energy to the plasma edge by heat conduction due to the Coulomb collisions of plasma particles, the plasma also emits energy by radiation. The latter process is based on the energy transport by electromagnetic waves and is not bound to matter. In plasma, a freely moving charge carrier emits then electromagnetic waves when it performs an accelerated motion in the form of change in speed or direction. This is for example the case when the charge carriers meet and their velocity is changed under the effect of the Coulomb force during flyby. For the emission of a hot plasma, the accelerated motion of electrons in the flyby to an ion is of interest. They emit so-called bremsstrahlung or X-ray bremsstrahlung[10] in the form of a continuous electromagnetic spectrum in the direction of the instantaneous velocity of the electron. The radiation power of electrons at the expense of their kinetic energy is equal to [1-7]:

$$P_{Br} \quad = \quad 5.4 \cdot 10^{-31} \cdot n_e \cdot n_i \cdot Z_i^2 \cdot (k \cdot T_e)^{1/2} = \quad c_{BR} \cdot n^2 \cdot T_e^{1/2} \tag{1-10},$$

with P_{Br}: radiation power by bremsstrahlung per unit volume of plasma

n_e, n_i: number of electrons or ions per unit volume of plasma ($n_e = n_i = n$)

Z_i: atomic number of plasma ions ($Z_i = 1$ for hydrogen plasmas)

k: Boltzmann constant = $1.38 \cdot 10^{-23}$ (J/K)

T_e: electron temperature (K)

c_{Br} = constant = $5.4 \cdot 10^{-31} \cdot Z_i^2 \cdot k^{1/2}$.

From Equation 1-10 it can be clearly seen that the Bremsstrahlung losses increase with Z squared. It is therefore important to achieve a very pure DT plasma and to avoid possible contamination with other ions.

From the fact that due to the Lorentz force the spiral paths of the charge carriers in the magnetic field constitute an accelerated motion, also here, electromagnetic waves

---

[10] The term X-ray bremsstrahlung comes from the fact that the mechanism is similar to that in X-ray tube, in which the electrons are slowed down in the entrance to the anode of the tube and emit X-rays.

are generated and emitted. This type of radiation of the plasma is called cyclotron or synchrotron radiation. It lies in a suitable wavelength ($\lambda = 1$ mm corresponds to a frequency f of about 10^{-11} 1/s), creating almost no loss and is well reabsorbed in the plasma again. The emission therefore takes place mainly from the plasma surface. It is also well reflected by metal walls and therefore partially get back into the plasma. The bremsstrahlung, in contrast, is much more penetrating than the synchrotron radiation. They come directly after emission by the electron from the plasma onto the wall of the plasma vessel and is absorbed there. It is therefore considered as the only inevitable radiation loss to the ignition condition.

1.5   Why Divertors?

1.5.1   How to Limit the Plasma Edge

As mentioned above, the heat and particle losses from the plasma occur only by a transport perpendicular to the magnetic field. On open field lines that intersect a wall material, however, the energy loss flows parallel to the magnetic field much faster. This leads to a strong plasma-wall interaction and consequently to a high wall load when coming in contact with the plasma.

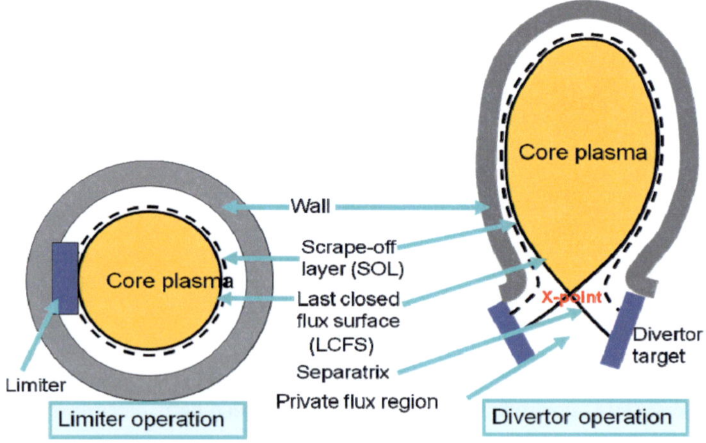

Figure 1-7: Limiter and divertor operations.

The limitation of the plasma edge through solid wall structures can be realized in different ways. A simple way is the use of a material limiter (Figure 1-7, left). It consists of plates which are directly brought into the hot plasma [1-14]. In this way, the last closed flux surface (LCFS) is defined. The magnetic field between de LCFS and the wall is known as scrape-off layer (SOL) [1-6]. The particles and heat which flow away perpendicularly through the LCFS, are in the SOL region mainly parallel to the magnetic field dissipated on the wall. The limiter has been used in previous experimental facilities (e.g. JET). However, it was found that sputtered atoms from the limiter itself (e.g. iron, nickel, chromium, oxygen) [1-15] due to the high load led to strong plasma impurities and thus energy losses in the plasma (see chapter 1.4.2).

1.5.2 Role and Functions of the Divertor

A better method is the magnetic plasma boundary, so-called divertor configuration [1-16] (Figure 1-7, right). Here, the limitation of the plasma is defined by a singularity in the magnetic geometry, the so-called X-point, which is generated by a magnetic quadrupole field. The surface magnetic flux passing through the X-point is called separatrix. Below the X-point a cold and high-density plasma region, so-called private flux region, is formed, which is separated from the plasma core. The divertor plasma configuration was intensively studied in the experimental reactor ASDEX (Axially Symmetric Divertor Experiment) in Garching in the early 70's. Aim of these experiments is to generate clean plasmas using divertors and to study the significance of the divertor for a future fusion power reactor. The breakthrough came in 1982 when a novel plasma state, the high-confinement regime (H-regime or H-mode), was discovered in the experiment "Divertor I". The H-regime develops independently of the type of heating only above a characteristic heating power threshold. The energy confinement time has doubled here, compared to the normal low-confinement regime (L-regime or L-mode). This means that with the help of the divertor, a very good plasma isolation resulting in clean plasmas has been achieved, making the divertor to the standard component of modern tokamaks. Accordingly, the large European Community Experiment JET in England (Figure 1-8) and the fusion experiment Doublet (DIII-D) in the U.S. were retrofitted, and the Japanese experiment JT-60 adapted to the divertor geometry of the ASDEX. Today's fusion experiments such as ASDEX-successor ASDEX Upgrade (AUG), TCV in Switzerland, KSTAR in South

Korea, EAST in China, and last but not least ITER (Figure 1-9) build on the divertor right from the outset.

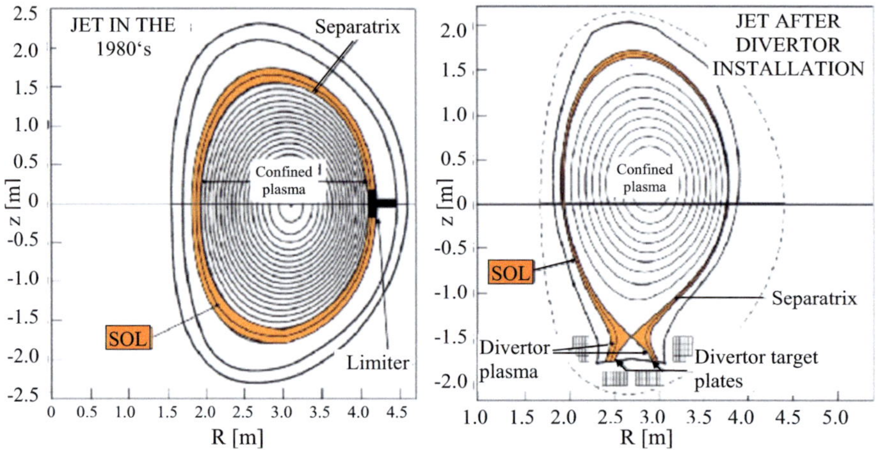

Figure 1-8: JET, the Joint European Torus, configurations before and after retrofit.

Consequently, the main function of the divertor is to remove most of the α-particle (fusion reaction ash), unburnt fuel, and eroded particles from the reactor. The latter are abraded from the first wall and have to be removed from the plasma, because they represent impurities that adversely affect the quality of the plasma. In general, maintaining a helium ash concentration below ~5–10 % is required in burning plasma. About 15 % of the total thermal power gained from the fusion reaction have to be mastered by the divertor, which results in a considerably high heat load of about 10 MW/m² on the relatively small divertor target surface, depending on the configuration and shape of the plasma. This energy fraction also plays a role in the total balance of the power station and, therefore, has to be used in an economically efficient manner, i.e. it has to be included in the power generation cycle.

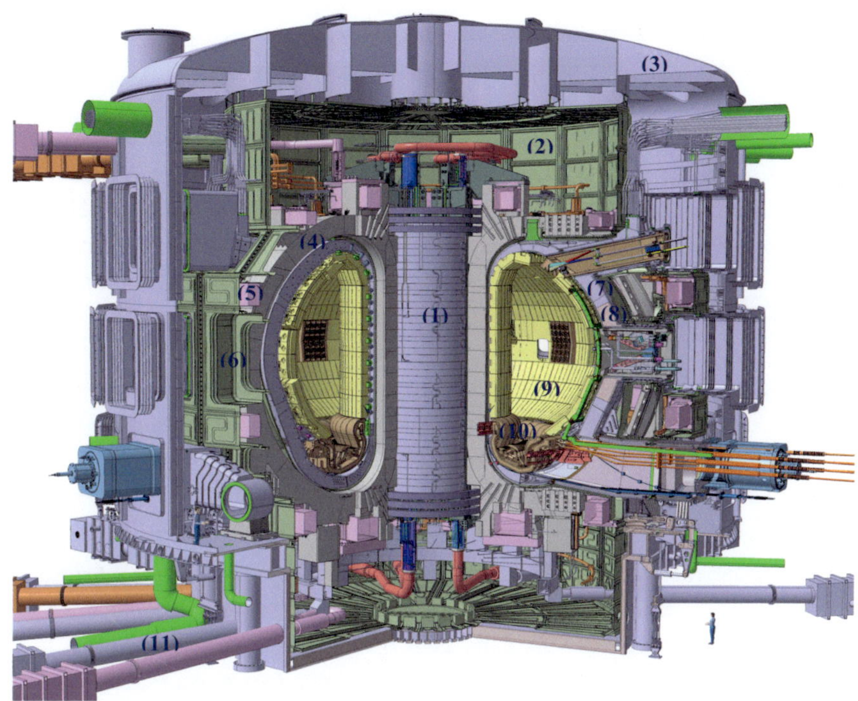

Figure 1-9: The ITER plasma machine. Fusion gain Q = 10, Fusion Power: 500 MW, Ohmic burn 300 to 500 s, Goal: Q = 5 for 3000s. Image courtesy of G. Janeschitz/ITER; http://www.iter.org/. Main components: (1) Central solenoid (n=6) (Nb3Sn), (2) Thermal shield (4 sub-assemblies), (3) Cryostat (Ø28 m x 29 m height), (4) Toroidal field coils (n=18) (Nb3Sn), (5) Poloidal Field Coils (n=6) (NbTi), (6) Correction coils (n=18) (NbTi), (7) Vacuum Vessel (9 sectors), (8) In-Vessel coils (2-VS & 27-ELM), (9) Blanket (440 modules), (10) Divertor (54 cassettes), (11) Feeders (31) (NbTi).

1.6   Tokamak Operation and Transient Phenomena Affecting the Divertor

In the tokamak the initial heating is generated by the transformer toroidal plasma current (resistive heating). Since the electrical resistance of the plasma decreases with increasing temperature, the plasma temperature cannot be increased above $kT \approx 1$ keV

($\approx$ 10 million K) with resistive heating. Therefore, an additional high-frequency or neutral particle heating is required to increase the plasma temperature. In the first kind, the energy of high frequency electromagnetic waves that are radiated into the plasma is absorbed by the plasma. In the latter, the heating is produced by accelerated neutralized deuterium particles, which are injected into the plasma and release their energy there through collisions. Furthermore, also the alpha particles with their initial energy of 3.5 MeV can heat up the plasma via collisions. This heating mechanism occurs, however, only above the ignition temperature of about 40 million Kelvin. Thereafter, in the ideal case, the alpha particles alone can cover all the exclusive radiative energy losses and maintain the ignition temperature of the plasma without additional heating. The alpha particles energy absorbed by the plasma corresponds in the ideal case 20 % of the total fusion energy and must be decoupled in the steady-state operation, resulting in a high wall loading of structures in particular for the divertor. During the stationary operation, it is also equally as necessary to continuously replace the burned fuel.

As described above, the necessary plasma current within the tokamak is maintained by means of a transformer, so that the screwing of the magnetic field lines is guaranteed. This works only under the constant increase of the magnetic flux in the transformer to prevent the decay of the plasma current. Due to the technical limit of the transformer, the tokamak operation must be interrupted after a certain time and the plasma must be ignited again. This is a so-called pulsed operation [1-17].

The stellarator generally allows steady-state operation, wherein the heating of the plasma in the initial phase and the refilling the spent fuel can be performed faster by means of the injection of fast neutral particles or pellets or cluster.

1.6.1 Edge-Localised-Modes (ELMs)

In the ASDEX divertor experiments it was recognized that a narrow transport barrier is formed at the plasma edge, resulting in the steep temperature and density gradients. In this barrier, the plasma turbulence being responsible for the poor thermal insulation is suppressed by shear flow almost completely. That is the key mechanism for improving the energy confinement time. As a result of good confinement, however, plasma edge instabilities, so-called Edge-Localised-Modes (ELMs), arise because the pressure gradient at the plasma edge quasi-periodically run at a stability limit. ELMs may lead to very high transient thermal loads on plasma facing components such as divertor target plates and present a serious danger to their life in the fusion power plant operation. Therefore, research is being done worldwide to prevent the ELMs formations. External resonant magnetic interference fields (RMP) [1-18] provide a method for suppression of boundary layer instabilities (ELMs) for future fusion reactors like ITER. Another promising method to achieve "ELM-free H-mode" shows for example a process combination of central heating and a bullet injection of small cryogenic deuterium pellets with addition of argon [1-19]. This causes a cooling of the plasma edge, without significantly increasing the contamination of the plasma.

1.6.2 Vertical Displacement Event (VDE)

To the present state of knowledge, a tokamak torus with vertically elongated cross-sectional shape (D form) is to be preferred, because it brings, in addition to the H-mode, a significant reduction of contamination of the plasma [1-6]. However, because of its non-circular cross-section the plasma has a tendency to an inherent instability and consequently effecting small vertical displacements. In practice, they can be controlled. However, large disturbances in the plasma, such as ELMs and disruption, affect the control loop and lead to a feed-back control error. The plasma can then move vertically up or down and in extreme cases to have contact with the vessel wall. This leads to large poloidal halo currents flowing locally from the plasma directly into the structure and from there back into the plasma. This leads to strong thermal and mechanical loads on the structure. This phenomenon, which is induced by the plasma crashes, is called a Vertical Displacement Event (VDE).

1.6.3 Plasma Disruption

A stable operating region of a tokamak is dictated by various parameters such as plasma density n, the weighted plasma pressure ß, and safety factor q. If one of these parameters lies outside of the critical limits, a sudden termination of plasma confinement can occur. Other instabilities, such as MARFE (Multifaceted Asymmetric Radiation From the Edge), a radiating thermal instability of the boundary layer, can lead to plasma disruptions, its causes and detailed processes are to date still not fully understood. During a disruption, a large part of the plasma kinetic energy $(1.3\text{–}7.5~\text{MJ/m}^2)$ is dissipated in a short time ($\sim 3\text{–}1.5$ ms) [1-20] on the first wall and in particular the divertor or limiter. This leads even with intact cooling to high thermal loads at the surface that it melts in part (thermal quench). It can also come to the evaporation of molten material. In addition, in a plasma disruption, eddy currents are induced in the plasma surrounding structures due to the breakdown of the poloidal magnetic field (current quench, $\sim$2 MJ/m^2, duration 10–50 ms [1-21]), which is generated by the plasma current. These lead to strong dynamic Lorentz forces in the plasma facing structures due to the existing toroidal magnetic field. As for ELMs suppression, here too, research is being conducted worldwide to eliminate the causes of disruption to avoid that risk to the limitation of lifetime and to the failure of the structure.

1.7 Objectives of this Work

The aim of this work is to develop a feasible divertor concept for use in a power plant to be built after ITER such as a demonstration reactor (DEMO). Developing a viable divertor concept is deemed to be an urgent task to meet the EU Fast Track scenario [1-22], where electricity production by fusion is to be achieved by 2030 and fusion power is to be commercialized by 2040.

This task is particularly challenging because of the wide range of requirements to be met, namely, the high incident peak heat flux, the blanket design with which the divertor has to be integrated, sputtering erosion of the plasma-facing material caused by the incident particles from the plasma, radiation effects on the properties of structural materials, and efficient recovery and conversion of a considerable fraction ($\sim$15 %) of the total fusion thermal power incident on the divertor.

After summarizing a literature research on the status of knowledge in chapter 2 including a review of the completed study EU PPCS and the initial divertor studies [1-23] in this framework, the following chapters 3 and 4 describe the path to the goal including the following objectives:

- Identification of the divertor heat loads (peak heat flux and distribution)
- Choice of suitable materials and coolant
- Positioning, toroidal segmentation, and outer design layout of a divertor cassette
- Conceptual design and thermal-hydraulic layout of a suitable heat transfer system able to remove the heat load
- Technological study on fabrication of divertor test mock-ups
- Design verification and proof of concept by tests.

2 State of Knowledge

The ultimate goal of the fusion program is the development of large-scale power plants for the production of electricity. The research and testing of the technology for nuclear fusion began in the late eighties at Karlsruhe Institute of Technology (KIT) Campus North (formerly Research Center Karlsruhe). One of the focuses is the development of plasma facing components, the so-called blanket and divertor, for an EU Demonstration Power Plant (DEMO). Thus, in 1995 at the EU reference concept selection, the KIT Helium Cooled Pebble Bed (HCPB) blanket concept next to the French Water Cooled Lithium Lead (WCLL) was successfully selected.

2.1 The EU Power Plant Conceptual Study (PPCS)

Later in 1999, an EU power plant conceptual study (PPCS) [2.1-1] was launched. One of the main objectives of this study is to achieve the highest possible efficiency, which contributes to a better economy of the plant, and thus makes the fusion power plant (FPP) environmentally friendly and attractive.

In the course of the EU PPCS three near-term (A, B and AB) and two advanced power plant models (C and D) (Table 2.1-1) were investigated. All models have an electrical capacity of 1500 megawatts, which was adopted for the comparison of various options, and are building type like ITER tokamak. The near-term models are based on limited extrapolations, both in physics and in technology, while more advanced ones use an advanced physics scenario combined with advanced blanket concepts. The plant models differ in their plasma physics, fusion power, as well as blanket and divertor technologies. Model A [2.1-2] utilizes a water-cooled lead–lithium (WCLL) blanket and a water-cooled divertor with a peak heat flux (PHF) of 15 MW/m^2. Model B [2.1-3] uses a He-cooled ceramics/beryllium pebble bed (HCPB) blanket and a He-cooled divertor concept (PHF 10MW/m^2). Model AB [2.1-4] uses a He-cooled lithium-lead (HCLL) blanket and a He-cooled divertor concept (PHF 10 MW/m^2). Model C [2.1-5] is based on a dual-coolant (DC) blanket (lead–lithium self-cooled bulk and He-cooled structures) and a He-cooled divertor (PHF 10 MW/m^2). Model D [2.1-6] employs a self-cooled lead–lithium (SCLL) blanket and lead–lithium-cooled divertor (PHF 5 MW/m^2). This shows that helium-cooled divertor designs are used in most of the EU plant models; it has also been proposed for the US ARIES-CS [2.1-7] reactor study.

	Model A near term	Model AB near term	Model B near term	Model C advanced	Model D very advanced
Blanket type Structure Coolant Breeder, function materials	WCLL Eurofer water PbLi	HCLL Eurofer helium PbLi	HCPB Eurofer helium Li-ceramics, Beryllium	Dual Coolant Eurofer PbLi & He ODS Eurofer, SiC$_f$/SiC	Self-cooled SiC$_f$/SiC PbLi
Reference divertor	Water cooled	Helium cooled	Helium cooled	Helium cooled	PbLi cooled
Maximum divertor PHF [MW/m^2]	15	10	10	10	5

Table 2.1-1: The EU PPCS plant models [2.1-1].

From these studies it appears, among other things, that the use of helium as a coolant for the blanket and in particular for the divertor is optimal solution for this requirement. In addition, helium exhibits a high level of security due to its chemical and neutronic inertness.

Basis for the conceptual design of the the He cooled divertor is the EU PPCS study of model C [2.1-5] including the incident plasma loads and reactor integration. Figure 2.1-1 shows a section through this reactor model. Both blankets and divertors are plasma facing components (PFCs), of which the divertor is located on the bottom of the reactor vessel. Its main function is to remove most of the α-particle (fusion reaction ash), unburnt fuel, and eroded particles from the reactor. The latter are abraded from the first wall and have to be removed from the plasma, because they represent impurities that adversely affect the quality of the plasma. In general, maintaining a helium ash concentration below ~5–10 % is required in burning plasma. About 15 % of the total thermal power gained from the fusion reaction have to be mastered by the divertor, which results in a considerably high heat load of about 10 MW/m^2 on the relatively small divertor target surface, depending on the configuration and shape of the plasma. This energy fraction also plays a role in the total balance of the power station and, therefore, has to be used in an economically efficient manner, i.e. it has to be included in the power generation cycle.

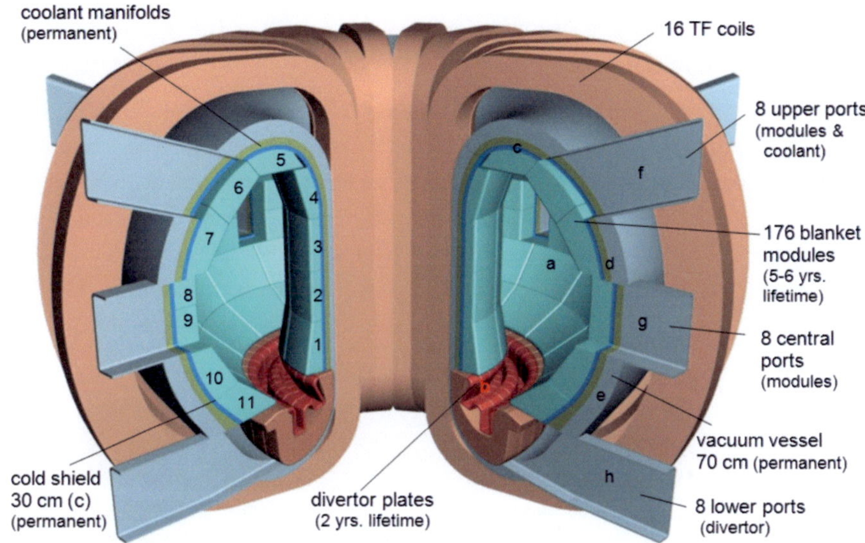

Figure 2.1-1: PPCS reactor model C [2.1-5] used as a basis for divertor study.

2.2 The ITER Divertor

The only existing example of an actual divertor design was the ITER divertor [2.2-1] (Figure 2.2-1) which is a water-cooled type. It operates at 4.2 MPa inlet pressure and relatively low temperature (100 °C to 126 °C at the outer vertical target (VT) and 127 °C to 141 °C at the inner VT), and at a low neutron flux [2.2-2–2.2-4] (see Table 3.4-1). Each plasma-facing component (PFC) of the divertor comprises a number of elements of 20–44 mm toroidal width with water coolant flowing in channels in the poloidal direction. The reference design for the strike point region (lower part of the VT with 10–20 MW/m^2 heat flux) uses carbon fibre composite (CFC) monoblock with an active metal cast (AMC$^®$) CFC/Cu joining, copper chromium zirconium (CuCrZr) heat sink, and a swirl tape insert in the coolant tube channel. The AMC$^®$ joint, which keys into the CFC, is obtained by casting pure Cu onto a laser-textured CFC with a Ti coating that aids wetting. The pure Cu is then joined to the Cu-alloy heat sink by brazing or hot isostatic pressing (HIP), with the additional option of electron beam (EB) welding in the case of flat tile geometry. For the upper part of the VT (5 MW/m^2 heat flux) the selected reference design employs tungsten tiles (10 mm × 10 mm × 10 mm) with a cast pure Cu interlayer, brazed or HIPed onto a CuCrZr structural material (heat sink).

The use of the high thermal conductivity CuCrZr heat sink enables high performance of the divertor, on the one hand. On the other hand, its embrittlement at the high neutron flux, as well as a reduction of fracture toughness at a neutron damage dose of 0.3 dpa (available data), especially at elevated coolant temperatures was reported in [2.2-5] and [2.2-6], respectively. This may well be considered a drawback that causes certain doubts in the applicability of such a material in an FPP environment under high neutron flux when operating at high coolant temperatures.

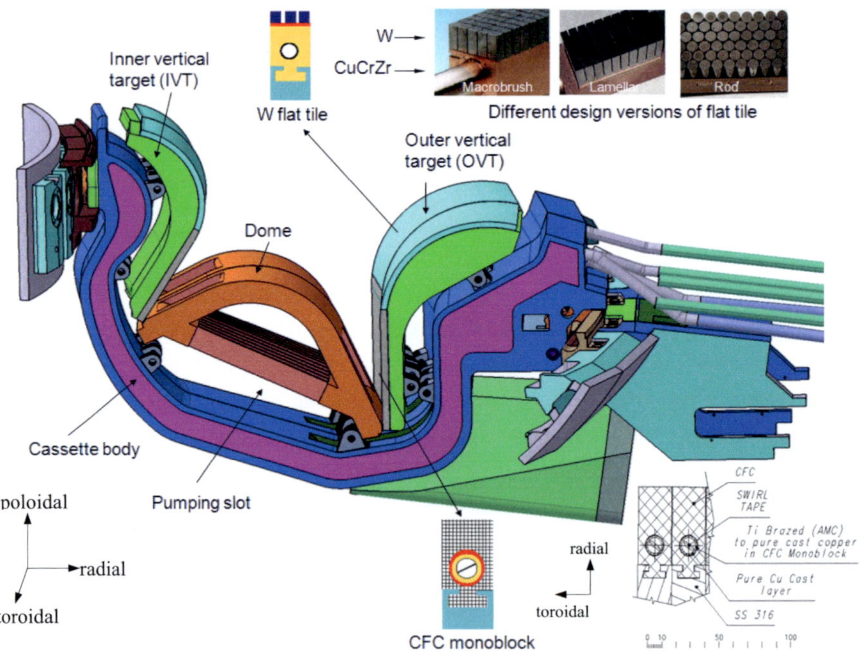

Figure 2.2-1: ITER divertor [2.2-1] (here: a central cassette) indicating the main components.

2.3 Review of the PPCS Divertor Design Studies

2.3.1 Water Cooled Divertor for PPCS-A

A water-cooled divertor (WCD) [2.3-1, 2.3-2] has been selected for the plant model PPCS-A. It is strongly based on the ITER divertor reference design [2.2-1] taking advantage of the limited extrapolation required from both the physics and technology developed and tested for ITER. An advantage of this divertor type is that the technology of water cooling circuits is well established; experience from water-cooled fission reactors (mainly PWR's) can be extrapolated to fusion reactor conditions. The choice of WCD also fulfils the PPCS requirement of using the same coolant throughout the reactor.

The initial WCD concept for PPCS-A assessed in 2001 [2.3-1] (Figure 2.3-1-i) uses a W-alloy monoblock (e.g. W 1.0 % (by weight) La_2O_3 (WL10), size ~20 mm radial x 18 mm toroidal), with a 3.5 mm thick sacrificial layer, a 2.5 mm deep lateral castellation for stress reduction, and an embedded CuCrZr water coolant tube (Ø11x1). The tube material was selected because of its superior fracture toughness compared to other Cu alloys. Similar to the ITER divertor, medium-temperature water (inlet temperature 140 °C, inlet pressure 4.2 MPa) is used and swirl tapes are placed within the tube to enhance the maximum acceptable critical heat flux. Oxygen Free High Conductivity (OFHC) Cu is used as a compliant layer inserted between the CuCrZr tube and the W-alloy monoblock. The thermo-mechanic analyses for a water coolant temperature as in the ITER divertor show that this concept can withstand a maximum heat flux of 15 MW/m^2. All temperatures and stresses are within the allowable limits.

A more advanced WCD conceptual design [2.3-2] (Figure 2.3-1-ii) was later introduced aiming at increasing the thermal efficiency by raising the water coolant outlet temperature to about 325 °C at 15.5 MPa pressure. It is based on the use of a series of poloidally-oriented EUROFER (the reduced activation steel developed in EU for fusion application) coolant pipes (Ø11x0.5) which allow to increase the water temperature up to PWR conditions to allow good heat conversion efficiency. Each of the pipes is surrounded by brazed W-alloy monoblocks and fixed on a common EUROFER back plate. A sacrificial 5.5 mm thick W layer is assumed. A swirl tape made of EUROFER is placed within the tube to promote turbulence. The temperature distribution was improved by including a compliance layer of a soft-graphite material

("Papyex" 0.1 mm thick) on each EUROFER tube and a thin layer of pyrolitic graphite partly deposited on the front inner surface of the W monoblock, which serves both as a heat flux repartitioning layer and a thermal barrier thereby reducing the maximum heat flux and the corresponding temperature gradients. Its thickness varies gradually from 0.075 mm in the front region down to zero in the lateral sides. The analytical results confirmed that this concept can withstand an incident surface heat flux of 15 MW/m^2. The use of heat flux repartition and the thermal barrier made it possible to achieve a safety margin of about 1.28 on the critical heat flux while maintaining the Eurofer structure temperature below the admissible limit of 550 °C. However, fabrication and irradiation issues for this design require future R&D to demonstrate the feasibility of such a conceptual proposal.

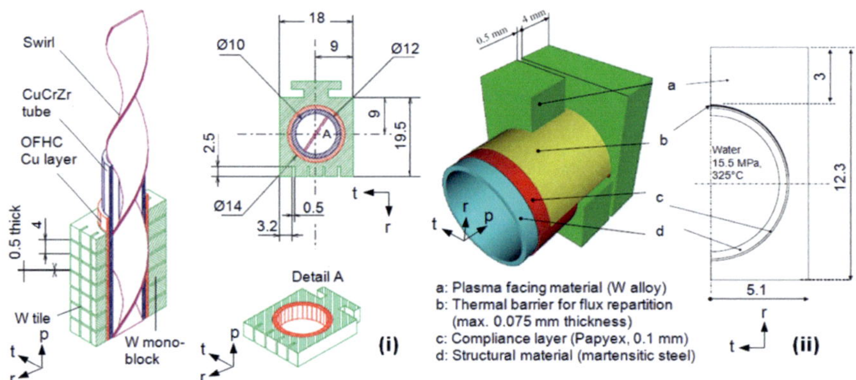

Figure 2.3-1: PPCS-A water-cooled divertor (q=15 MW/m^2 required), an extrapolation of the ITER design: (i) W/CuCrZr concept [2.3-1], (ii) Concept with RAFM steel heat sink [2.3-2]. Coordinates in r = radial, p = poloidal, t = toroidal.

2.3.2 Liquid Metal Cooled Divertor for PPCS-D

A forced-convection lead-lithium (Pb17Li)-cooled divertor design [2.3-3] (Figure 2.3-2) was chosen for the plant model PPCS-D. The divertor has to handle a maximum peak heat flux of 5 MW/m^2. The divertor target plate consists of a number of poloidally-oriented silicon carbide-silicon carbide composite (SiC$_f$/SiC) square tubes (Figure 2.3-2, right), with a 5.5 mm thick sacrificial layer of tungsten alloy armor. A

"T flow separator" is inserted in each tube which assumes a comb form in the region nearest to the plasma, thereby creating toroidal channels. The lead/lithium eutectic Pb17Li flows poloidally in one-half of the tube (serving as an inlet header), then it is forced to pass through the short toroidal channels to cool the high flux region through a very short path, and finally it is routed back to the other side of the poloidal tube which serves as an outlet header. The channel dimensions and the liquid metal velocity in the different regions of the divertor are varied in order to adapt them to different heat loads. The poloidal tubes in the HHF region are 30 mm deep and 28 mm wide; the depth of the toroidal channels is 1.4 mm. Each divertor segment accommodates 22 poloidal tubes (only two are shown in the top right of the figure), each of which forms 32 toroidal channels (bottom right in the figure). The velocity of the Pb17Li in the toroidal direction ranges from 1.5 m/s in the front to 1 m/s in the rear. The thickness of the poloidal SiC_f/SiC tube varies from 1 mm in the region near the plasma (to lower the temperature gradients and stresses) up to 2 mm in the back and side walls (required to withstand the internal pressure).

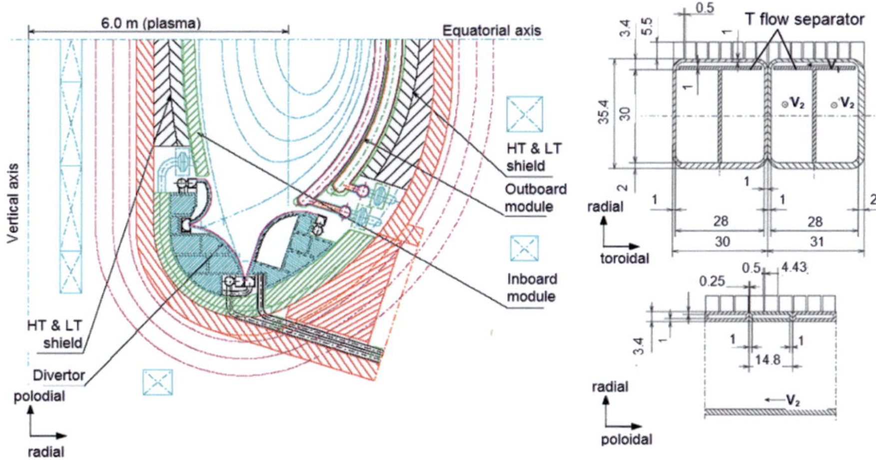

Figure 2.3-2: PPCS-D liquid metal-cooled divertor [2.3-3] (q = 5 MW/m2 required). Left: cutout of a 7.5° sector, right: toroidal-radial and poloidal-radial cross-sections of the divertor target plate in the HHF region.

For the thermo-mechanical analyses of the PPCS-D divertor it was assumed a surface heat flux of 5 MW/m^2 with an inlet Pb17Li temperature of 600 °C, the calculated maximum temperatures at the channel outlet (W 1288 °C, SiC$_f$/SiC 1016 °C) are within acceptable engineering limits. These calculations were based on large extrapolations for the assumed material physical properties (e.g., SiC$_f$/SiC thermal conductivity of 20 W/mK), which, together with other open issues such as joining technology, neutron irradiation effect, and magnetohydrodynamic (MHD), require significant and long term R&D.

2.3.3 Helium Cooled Divertor for PPCS Models AB, B, and C

Helium cooling offers several advantages including chemical and neutronic inertness and the ability to operate at higher temperatures and lower pressures than those required for water cooling. The drawback is its comparatively low heat exchange capability as well as the considerably large pumping power. The former can be enhanced in various ways, e.g. by promoting turbulence and/or by increasing the solid/fluid interface area.

A helium cooled divertor (HCD) has been selected for the models PPCS-B, C, and later defined model AB (after 2004). It simplifies the balance of plant since the same coolant is used for all internal components, thereby allowing the power conversion systems to be well integrated. Additionally, for PPCS-B it eliminates the risk of hydrogen formation from the water–beryllium reaction in the event of an accident. HCD investigations began in 1999 within the framework of the EU power plant availability study (PPA) and the following first stage of PPCS in 2000. Several initial concepts [2.3-4, 2.3-5] had been considered. Helium gas operating pressures of 10–14 MPa with an inlet temperature of about 600 °C are typically assumed.

The unconventional design [2.3-4] (1999) (Figure 2.3-3, left) uses a porous medium heat exchanger and can accommodate a peak heat flux of 5–6 MW/m^2. The porous medium provides a high surface area-to-volume ratio favourable for the heat transfer enhancement; it also provides an irregular coolant flow pattern favourable for turbulent mixing. This design utilizes helium at 8 MPa with an inlet temperature of ~630 °C and an exit temperature of 800 °C, which is compatible with the operating temperature of the structural material (TZM[11] or W alloy). The helium is forced through a slot at the top of the coolant inlet tube into a circular porous wick layer (porosity ~40 %), flows sideward through the porous layer before exiting through a

---

[11] Molybdenum alloy with 0.5% Ti, 0.08% Zr, and 0.04% C.

bottom slot of the outlet tube. The typical effective heat transfer coefficient (HTC) (see the definition in equation 3.7-2) is about 20 kW/m²K at a maximum He velocity 140 m/s and a pressure loss of 0.45 MPa for 1 m target plate length.

The basic cooling principle behind the porous medium design was adapted to develop a simple slot concept [2.3-5] (Figure 2.3-3, centre) which relies on the heat transfer capability of the helium flowing through a narrow peripheral gap of 0.1 to 0.2 mm rather than through a porous medium. This approach simplifies the design and manufacturing of the coolant channel system by omitting the porous medium. In this study, the following helium coolant parameters were used: Inlet/outlet temperatures of 600/800 °C, 14 MPa pressure and a mass flow rate of 0.17 kg/s. This study yielded a typical effective HTC of about 14 kW/m²K at and a maximum He velocity 75 m/s and a pressure loss of 0.14 MPa for 1 m target plate length. The same heat flux level could be reached by using Multi-channel and Eccentric Swirl concepts in which the HTC is primarily enhanced by increasing the coolant velocity on the heated side of the coolant channel. The modified slot concept (2001) [2.3-6] (Figure 2.3-3, right) increased the heat flux limit to about 10 MW/m². It uses either a narrow peripheral gap of 0.1 mm thickness to increase the coolant velocity upon exiting the inlet channel or a pin array (with a larger peripheral gap of about 1 mm), through which the coolant passes before flowing into the outlet channel. This study yielded a maximum local HTC of about 60 kW/m²K at a maximum velocity of 200 m/s (mass flow rate per channel ~0.2 kg/s, pressure loss ~0.1 MPa per 1 m target length).

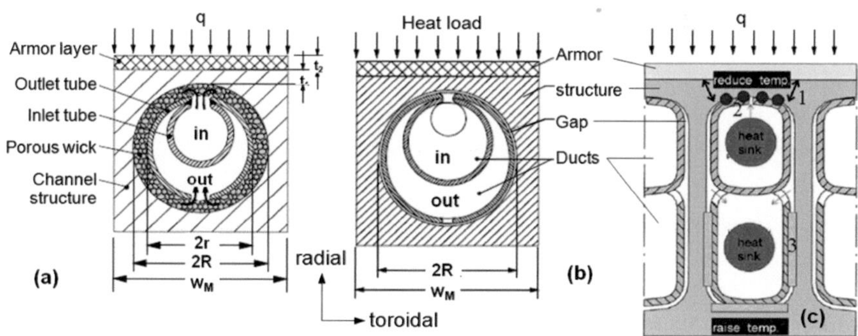

Figure 2.3-3: Some initial HCD designs: (a) porous medium concept [2.3-4] (q=5 MW/m²), reference dimensions [mm]: r=11, R=14, w_M=36, t₁=t₂=3; (b) simple slot concept [2.3-5] (q=5 MW/m²); (c) modified slot principle [2.3-6] (q=10 MW/m²): 1 reducing conduction paths, 2 maximizing htc e.g. by pin array, 3 maximizing isolation.

3   Detailed Design of a Helium Cooled Divertor for a Fusion Power Plant

3.1   Segmentation and Positioning of the Divertor in a Reactor

The divertor is toroidally divided into cassettes, e.g. 48 cassettes of 7.5° each for the PPCS-C (Figure 3.1-1), for easier handling and maintenance. It is essentially composed of the thermally highly loaded target plates, the dome and wings that contain openings for removing the particles by vacuum pumps, and the main structure or bulk which houses the manifolds for the coolant and, at the same time, serves as neutron shielding for the superconducting magnets behind it. Its position in the reactor depends on the configuration of the plasma-supporting magnetic field. It can be accommodated at the lowest and/or highest position of the vacuum vessel (the latter is indispensable in case of a double null plasma configuration). Together with the blanket, it forms a closed lateral surface or enclosure around the plasma.

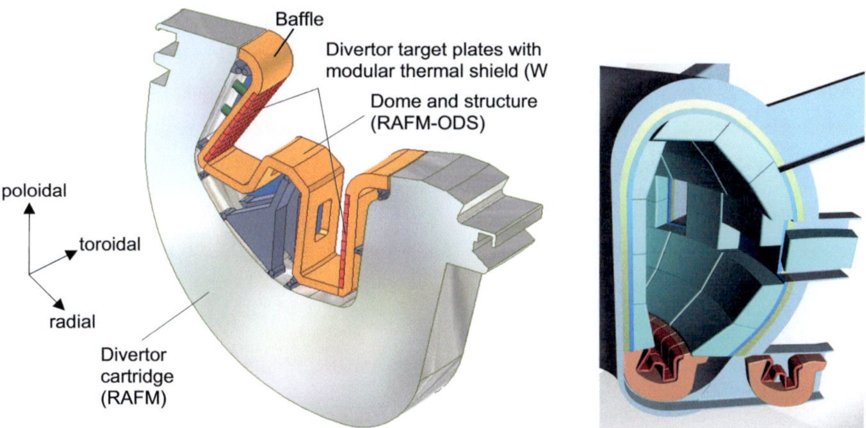

Figure 3.1-1: Basic design of a DEMO divertor cassette [2.1-5]. Right: replacement scheme.

The plasma-facing target plates are preferably made of tungsten (in ITER, tungsten monoblock with copper chromium zirconium inserts as heat sink shall be used) with a sacrificial layer of about 2–3 mm thickness. Regarding the choice of material see the following sections.

The target plates are positioned under a certain angle to the extension of the SOL magnetic field lines (see below), along which the α-particles with high kinetic energy and additional plasma heating energy are led to the targets. This causes a surface erosion of the target plates (therefore, the expression "sacrificial layer" is used), which is why the divertor must be exchanged frequently. Presumably, the target plates will reach a service life of 1–2 years before they will have to be exchanged. Furthermore, the divertor is exposed to a shower of neutrons which cause an additional volumetric heating in its body. For example, approximately 22 % of the total heat load of the outboard target plates is due to neutron heating. The supporting structures and the divertor cassette bodies are made from stainless steel (austenitic steel 316L for ITER or the reference ferritic-martensitic steel EUROFER/ODS EUROFER for DEMO). More details about the construction of ITER divertor can be found in [3.1-1].

In general, vertical target plates help optimally pushing neutrals towards the private flux region (chapter 1.5.1). According to practical guidelines in [3.1-2], the position of the vertical plate, however, has strong influence on the surface heat load of the plate itself and the direction of the reflected neutrals. A target plate perpendicular to the flux lines for example would push neutrals towards the X-point and thus causing X-point MARFE which may destroy the plasma. To achieve a compromise between an acceptable surface heat load on the target plate and a reasonable neutral density for pumping, the following guidelines [3.1-2] (Figure 3.1-2) are recommended:

a.   The vertical target angle is to be adapted that the maximum heat flux (during transient off-normal events) does not exceed 15–20 MW/m^2. The experience value for the poloidal angle of the target plate is approximately 15–25°, which corresponds to a total angle of 1–2°.

b.   For a given plasma configuration, the strike points of the impacting SOL should lie only within the vertical plate zone.

c.   The normal from the intersection of the 3-cm-SOL flux line from the plasma edge with the vertical target surface should not come into the private flux region (P) higher than the dome.

These guidelines ensure that the most of the recycling neutral fluxes are under the dome. They also define the necessary space between X-point and vacuum vessel for a functioning divertor.

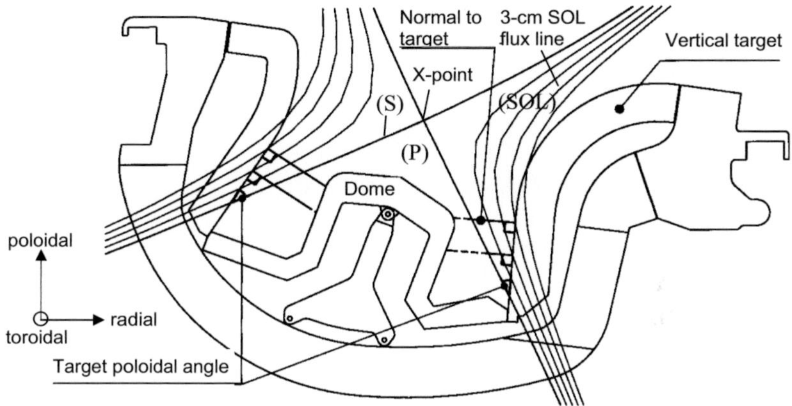

Figure 3.1-2: Sketch relating to guidelines for divertor geometry design [3.1-2]; P = private flux region, S = separatrix, SOL = scrape-off layer.

3.2 Functional Design Requirements

o   Resisting a peak heat flux of 10 MW/m² with an average of 5 MW/m².

o   A tolerance for the moving position of the peak heat flux is to be taken into account in a range of 40 cm at the lower end of the target plate.

o   An average neutron wall load of about 1.7 MW/m².

o   The divertor shall be designed for a lifetime of about ~1–2 years, within those it has to survive 100–1000 startup and shutdown thermal cycles, most of which are hot shutdown[12] (pulsed mode) or hot standby (emergency shutdown), and very few are cold shutdown (e.g. for maintenance and component replacement).

---

[12] That condition is when the reactor is scrammed, the generator is tripped but reactor coolant temperature is maintained by decay heat and/or reactor coolant pump heat input [3.2-1].

- o The divertor is to operate with helium coolant. The divertor heat is to be used for electricity production by integrating the divertor coolant into the power conversion system to maximize the net reactor efficiency[13].
- o Based on the PPCS strategy it is assumed that VDE can be avoided and ELMs suppressed. However, 10–100 full power disruptions should be taken into account as abnormal load assuming a current decay time during disruption of about 50 ms.

3.3   Identifying the Heat Dissipation System

As schematically shown in Figure 3.3-1, the incident heat load on the plasma facing surface of the target plate must be dissipated through its first wall by the cooling segment beneath it. The first wall, which accommodates a high temperature and a large temperature gradient ΔT due to the high surface heat flux $q_{surface}$ and volumetric heat generation q_{vol} has to fulfill the functions of a protective and sacrificial layer and a structure at the same time. The temperature level depends on the heat sink temperature in the cooling segment, the heat-transfer mechanism at the coolant/solid wall interface, and the temperature gradient in the wall itself dictated by the thermal conductivity of the wall material. Suitably, the first wall may consist of two separate layers of materials that meet the different requirements. In a heat transfer design they are to be bonded together in order to achieve at optimum heat transmission.

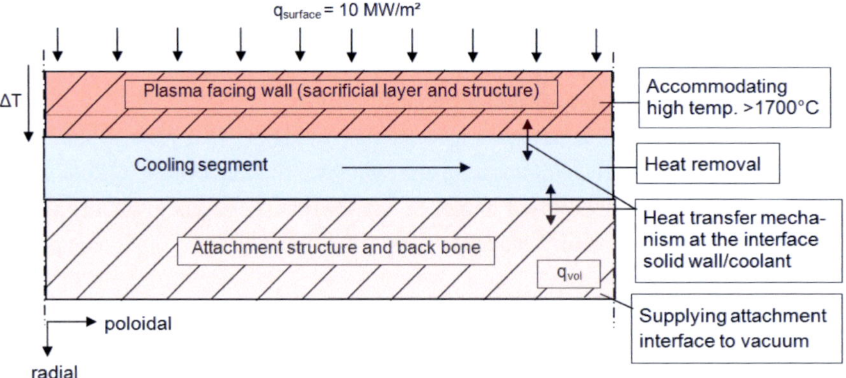

Figure 3.3-1: Sketch relating to the heat transmission problem of the divertor.

---

[13] The ratio between the electrical power output to the grid and the fusion power.

As a result, the following key design priorities lie in the choice of materials, the thermal hydraulic balancing, and not least in the thermo-mechanical check of the structure for its integrity. A modular design principle (Figure 3.3-2) has proven to be suitable design to minimize thermal stresses [3.9-2].

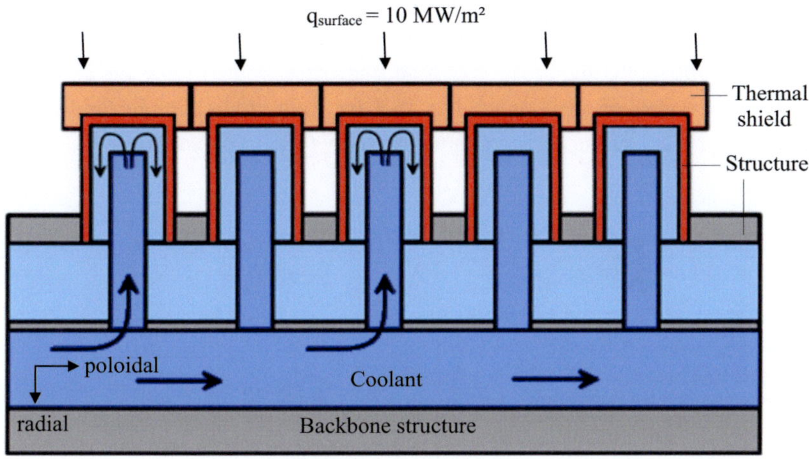

Figure 3.3-2: Modular design principle.

3.4   General Choice of Divertor Materials

In general, the plasma vessel is strongly exposed to plasma particles, neutrons and electromagnetic radiation. Charged and neutral particles contribute mainly to the plasma facing vessel surface, leading to physical and/or chemical sputtering. Therefore, the main determining factor for the choice of plasma facing materials (PFM) is the erosion lifetime. Also tritium retention in the bulk material or tritium trapping due to co-deposition of tritium with eroded material; in particular carbon is an important aspect hereby. Since the amount of tritium uptake in tungsten is small (ten times lower than C or Be) its use in areas of high neutral flux will help reduce the tritium inventory in the vessel.

Due to the high heat load and the plasma bombardment with a high incident particle flux of up to $10^{24}/m^2s$ anticipated for a power plant (Table 3.4-1), a sputter-resistant material for the divertor is required in order to keep the erosion of the divertor targets as low as possible because the radiation losses of the plasma increase with the square of the effective atomic number Z_{eff}[14] (see also chapter 1.4.2). Sputtering takes place when surface atoms of a solid receive sufficient impact energy by ions or atoms to exceed the surface binding energy so that they are removed as a result. There is also another mechanism, the so-called self-sputtering, that some of the metal atoms are sputtered off the target, being themselves ionized and then return to the target to knock off still more atoms. In general there is threshold energy for the occurrence of sputtering, below which no sputtering occurs (Table 3.4-2). For high Z materials the sputter yield is smaller than for light atoms like He or H, but self-sputtering is very strong above certain ion energy (e.g. 70 eV for W).

	ITER		Fusion Power Plant	
Major plasma radius [m]	6.2		7.5 – 9.6 (for ~1500 MWe)	
Neutron wall load [MW/m^2]	FW	Divertor	FW	Divertor
- average	0.56	~0.37 / 0.47[a]	~2.4	1.7
- max.	0.78	~0.38 / 0.49[a]	~3.1	~2.1
Peak particle flux [10^{23}/m^{2}s]	0.01	~10	0.02	~10
Max. surface heat flux [MW/m^2]	0.25–0.5	10[b]	0.5	10
Av. neutron fluence [MWa/m^2]	~0.3–0.5	~0.15	~10	~3–4
		(w/o replacement)		(2 years cycle)
No. of cycles	30000	(10000?)	< 1000	< 1000
Pulse length [s]	400, 1000–3000 advanced scenarios, ~1200 long dwell		steady state or long pulses (e.g. 10000 and short dwell)	
Blanket	o No tritium production		o Tritium production and extraction	
	o Water cooling		o He cooling	
	o Low coolant temperature (no electricity production)		o Higher temperatures for electricity production	
			o High shielding capability	
Divertor	o "Cold" divertor		o Divertor integrated in the power generation system (divertor heat ~15 % of the fusion thermal power)	
	o Water cooling			
Availability:	10 %		> 70–75 %	

[a] depending on scenario, [b] slow transients: 20 MW/m^2 lasting 10 s, 10 % frequency

Table 3.4-1: Comparison between major requirements for ITER and first generation reactor [2.1-1, 2.1-5, 2.2-2–2.2-4].

---

[14] The effective ionic charge $Z_{\text{eff}} = \Sigma_i\, n_i.Z_i^2\, /\, \Sigma_i\, n_i.Z_i$ is a means to assess the impurity content of a fusion plasma [3.4-1].

Ion →	H	D	T	^{4}He	self
Be	27	24	28	33	–
Graphite	10	10	13	16	30
Ti	44	36	28	22	41
Fe	64	40	37	35	35
Mo	164	86	50	39	54
W	400	175	140	100	70

Table 3.4-2: Sputtering threshold energy for target materials at different ion in eV [3.4-2, 3.4-3].

Low-Z elements C and Be were originally proposed as wall or divertor materials because of their good plasma compatibility. In recent experiments, however, they show relatively high sputtering under bombardment by hydrogen ions, so that they appear to be not suitable as a wall material. One advantage of carbon, however, is that it has no liquid phase and thus should have a more favorable behavior during intense thermal shock loads. But, by far the most significant disadvantage of carbon is the co-deposition of tritium with eroded carbon. An alternative to low-Z divertor materials is tungsten. The sputtering rates by hydrogen are much lower and the thermal properties are also suitable. Tungsten has thus the following advantageous properties for the divertor design application: high melting point, high heat conductivity, low sputtering rate, and low activation. However, because of its high atomic number, the concentration of tungsten in the plasma must be strongly restricted. Recent estimates for ITER result in a maximum allowed tungsten concentration of some 10 ppm [1-19].

For ITER tungsten (W) has been selected for most of divertor surfaces but Carbon Fiber Composite (CFC) is still used for the highest loaded target section. CFC has very good thermal shock resistance but some disadvantages in sputtering and tritium retention (large inventory). For DEMO tungsten has been selected for the whole divertor surface, whereby limiting thermal loads and thermal shocks for the divertor design is required.

One of the critical points for the structural material properties is the neutron exposure. The 14-MeV fusion fast neutrons have a considerable range of penetration through matter. They penetrate the first walls and bodies of blanket and divertor and release their energy through collisions. This leads to the displacement of lattice atoms and nuclear transmutation causing swelling, creep and embrittlement of the material. Therefore, the objective of the development of materials is that, in addition to achieving the highest possible resistance to such loads also reaching low-activation property of the materials, i.e. the activation level should decay as fast as possible.

Following criteria for the selection of the divertor functional and structural materials have to be met:

- low sputter yield, i.e. high sputtering threshold energy for D, T, He,
- low self-sputtering,
- low tritium retention,
- low activation[15] (Figure 3.4-1),
- high thermal shock resistance,
- high thermal conductivity,
- low thermal expansion coefficient,
- high strength (especially for structural materials).

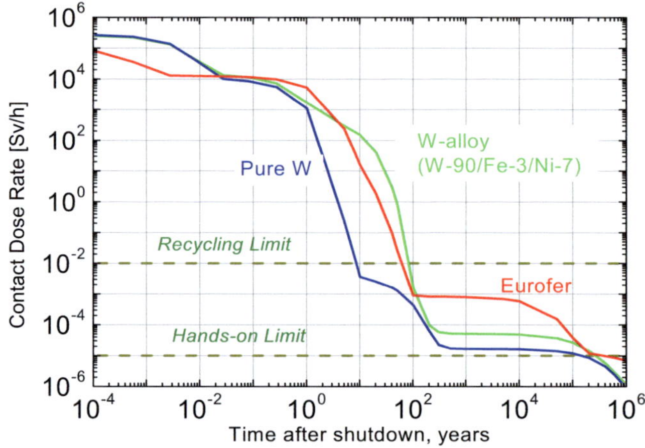

Figure 3.4-1: Contact dose rate of W, Densimet, and EUROFER after 2 FPY irradiation in divertor of the EU PPCS plant model B (HCPB) (chapter 2.1). Image courtesy of U. Fischer, KIT. More results for PPCS model AB (HCLL) see [3.4-4].

---

[15] There is no exact definition of a low-activation material. A value of 10 µSv/h is given in [3.4-5] as a realistic limit for hands on. As fuzzy criterion, the contact dose rate is used that should be under the hands-on limit over a period of about 100–200 years. Currently there is no material which strictly meets this strong criterion.

3.5 Functional Design Specifications

The above analysis of the constraints leads to the following functional design specifications:

- The divertor is divided into 48 cassettes to facilitate remote handling.
- The poloidal length of the target plate is 1 m, the length of the baffle (Figure 3.1-1) is 0.5 m.
- The outboard target plate is poloidally inclined by $10°$ relative to the strike plane to reduce the heat load on the surface. This value may be larger for the inboard target.
- The average heat load will be about 5 MW/m², the peak heat load 10 MW/m². The peak will be moving along the target plate in a range of 40 cm. The heat flux profile given in [3.7-1] will be assumed as working hypothesis for this study.
- Tungsten has been selected as armour or tile material because it has the best sputtering resistance and thermo-physical properties of all material candidates.
- For the structure directly underneath the tiles tungsten alloy, currently W 1.0 % (by weight) La_2O_3 (WL10), is employed with a relatively high thermal conductivity.
- The sacrificial layer on the target plates is assumed to be 5 mm thick (minimum 3 mm depending on the heat flux), which should be sufficient for a lifetime of about 2 years.
- A modular design in form of finger modules instead of large plate structures is preferred to reduce the thermal stresses (see chapter 3.3).
- The tungsten tile should be attached separately for reasons of containment integrity against crack growth.
- The basic structure is to consist of ODS steel. This is subject to the condition that a solution can be found for the transition joint between tungsten alloy and steel due to their large mismatch in thermal expansion coefficient (see details in chapter 3.6.1).
- To design the joints between the divertor components that fulfill the functions of withstanding the thermo-cyclic loadings and stopping the crack growth

introduced from the plasma-facing side to maintain the integrity of the structures.

- Transport of the coolant is to be realized as close as possible to the target plates in order to keep the maximum temperature of the structure as low as possible.

- Short heat conduction paths from the plasma facing side to the cooling surface must be ensured to keep the maximum temperature of the structure below the recrystallization temperature of the structure material.

- To achieve high heat transfer coefficients while keeping the coolant mass flow rate and, thus, the pressure loss as well as the pumping power as low as possible. The pumping power due to the pressure loss should not exceed 10 % of the thermal energy gain.

- To keep the divertor operating temperature window at the lower boundary higher than the ductile-brittle transition temperature and at the upper boundary lower than the recrystallization temperature of the structural part of tungsten alloy (see details in chapter 3.6.1).

- The concept should be feasible for manufacturing in a mass production process due to the large quantity of finger units required (> 300,000). Promising methods for producing tungsten parts are powder injection molding and deep drawing (see details in chapter 4.3).

3.6   Description of the Reference Design

Originally, some design ideas based on different kinds of heat transfer promoter have emerged. Figure 3.6-1 shows two promising design options HEMJ (He-cooled modular divertor with jet cooling) (left) and HEMS (He-cooled modular divertor with slot array) (right) using different heat transfer mechanisms. In the HEMS design [3.6-1], a tungsten slot array is used to enhance heat transfer at the bottom of the thimble by brazing it to the cooling surface, thereby increasing the heat transfer capacity with a predicted average HTC of 21 kW/m^{2}K. But on the other hand the additional supporting back plate of the slot-array results in a disadvantageously additional temperature gradient of about 100 K/mm. Furthermore, this design option brings a higher pressure loss than HEMJ as an experimental result has shown (see chapter 3.9-2). The HEMJ design [3.6-2, 3.6-3] is based on direct jet-to-wall impingement cooling with multiple helium jets. These are generated by a steel cartridge carrying an array of small jet holes, which is placed concentrically inside the thimble. The HEMJ reaches a predicted average HTC of 31 kW/m^{2}K (see more details in sections below). Although both designs are capable of withstanding an incident heat flux of 10 MW/m^2, the HEMJ design has been defined as a reference for its simple design and manufacturing. In the following chapters, detailed design and associated R&D work on this concept are described.

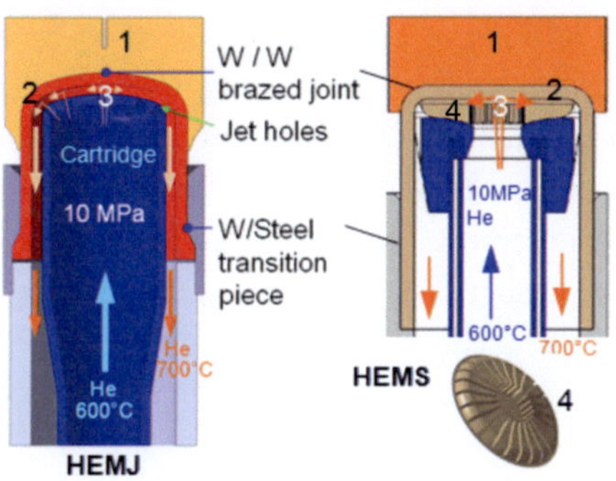

Figure 3.6-1: Modular He-cooled divertor designs HEMJ [3.6-2, 3.6-3] (left) and HEMS [3.6-1] (right). Both designs handle q=10 MW/m^2.

3.6.1 Construction

As illustrated in Figure 3.1-1, the main components of the divertor are the thermally highly loaded target plates, the dome with the opening for removing plasma impurities by vacuum pumps, and the main structure or bulk with the manifolds for the coolant. The surface of the target plates is provided with a thermal shield made of heat-resistant material tungsten, which possess favourable properties, e.g. high melting point, large thermal conductivity, and a low sputtering rate. To lower the thermal stresses, the tungsten armour layer is segmented to a size smaller than 20 mm. A hexagonal form of small segments allows a higher packing density for heat dissipation.

Today's reference concept, called He-cooled Modular divertor with Jet cooling (HEMJ) [3.6-3], is based on a modular design of small tungsten-based cooling fingers

(Figure 3.6-2). Such a modular design helps reduce thermal stresses. Therefore, each finger consists of a small hexagonal tile made of pure tungsten (18 mm width over flat and 5 mm thick) as thermal shield and sacrificial layer which is brazed to a thimble (Ø15 x 1 mm) made of tungsten alloy. Currently W 1.0 % (by weight) La_2O_3 (WL10) is preferred as thimble material because of its favorable property for·the machining. The reason for the separation of these two components (tile and thimble) is that the cracks initiated from the tile surface to be stopped at the interface. The tungsten finger units themselves are then connected to the support structure made of ODS steel (e.g. an advanced ODS EUROFER or a ferrite version of it) by means of brazing. To compensate for the large mismatch between tungsten and steel, a transition piece is required. The current solution uses a conical steel ring and a copper-based alloy as brazing material. The conical form of the joint serves as an interlock against flying away of the thimble. The divertor finger is cooled by multiple helium jets at 10 MPa and 600 °C impinging onto the heated wall of the thimble. The inlet and outlet temperatures of the helium coolant are restricted by the ductile-brittle transition temperature (DBTT) of irradiated WL10 (600 °C assumed) and the creep rupture strength of the ODS steel structure, respectively (Figure 3.8-21). The helium jets are generated by an array of small jet holes (Ø 0.6 mm) located at the top of a cartridge made of ODS Eurofer. The cartridge itself carrying the jet holes is placed concentrically inside the thimble. The number, diameter, and arrangement of the jet holes as well as the jet-to-wall distance are important parameters. The results of the Computational Fluid Dynamics (CFD) parametric study [3.8-1] show that the jet-to-wall distance (within the design range of 0.6 – 1.2 mm) has no excessive influence on the divertor performance. On the other hand, the jet hole diameter has a substantially larger influence on the divertor performance and the pressure losses. The following geometry was found suitable: 24 holes Ø0.6 mm and 1 center hole Ø1 mm, jet-to-wall spacing 0.9 mm.

For the nominal case with 10 MW/m^2 surface heat flux and 6.8 g/s mass flow rate (MFR), the calculation results yield a maximum tile temperature of 1700 °C which is well below the design limit of 2500 °C (Figure 3.8-21). The maximum thimble temperature amounts to about 1170 °C which is below the permissible value of 1300 °C assumed as recrystallization temperature of irradiated WL10. The calculated pressure loss (Δp) of 0.12 MPa seems to be overestimated, compared to the measured values of about 0.10 MPa from the earlier gas puffing experiments in Efremov [3.8-3]. These experiments were based on a reversed heat flux principle. A maximum divertor performance of up to about 12 MW/m^2 for the HEMJ design was found in these experiments at a nominal MFR of 6.8 g/s.

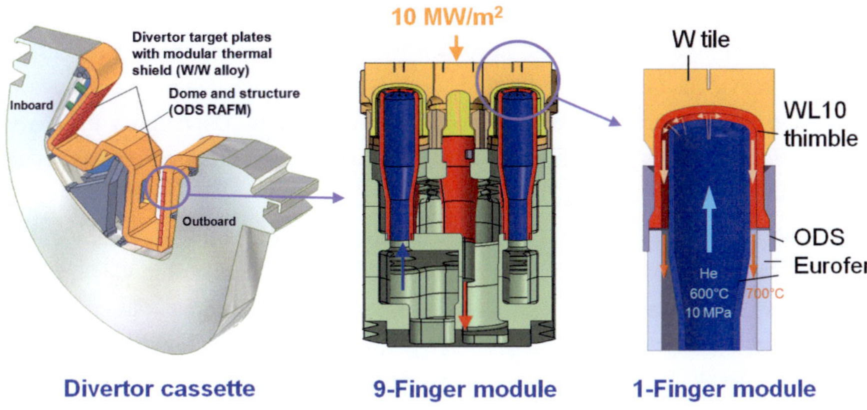

Figure 3.6-2: The reference modular design.

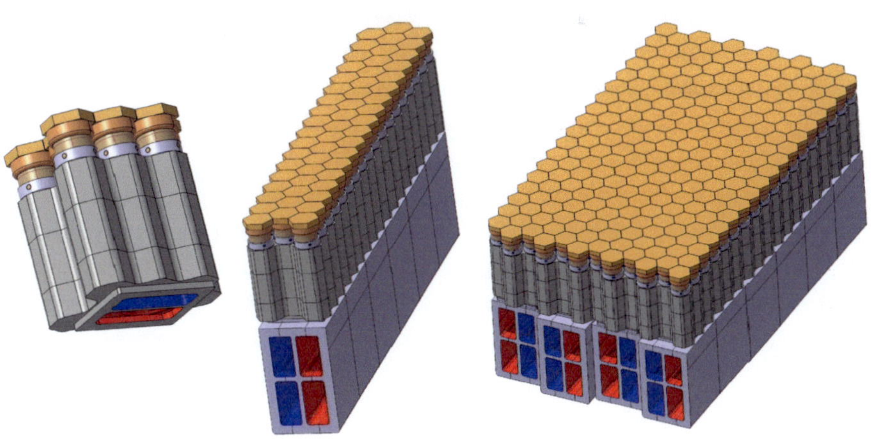

Figure 3.6-3: Modular structure of the divertor design.

3.6.2 Component Materials

Tungsten tile: The requirements for high resistance of the protective armor material against high heat flux (HHF) and sputtering erosion caused by the incident particle flux lead to the choice of tungsten as the most promising divertor material. It offers advantages in high melting point, high thermal conductivity, low thermal expansion and low-activation. On the other hand, it has high hardness and high brittleness, which is disadvantageous for the mechanical manufacturing of parts. The tungsten tiles have no structural function. A sacrificial layer of 2–3 mm is foreseen for an estimated service life of about 1–2 years.

Thimble of tungsten alloy: The materials should present both good thermal and mechanical properties to minimize stresses and temperature gradients in the high flux region at the plasma side and absolve its structural function. The operating temperature window of the W alloys structures is restricted at the lower boundary by the ductile-brittle transition temperature (DBTT), below which it loses its ductility associated with a suddenly occurring material failure. At the upper boundary, the temperature window is limited by the recrystallisation temperature (RCT), above which tungsten loses its strength due to grain coarsening. Generally, the DBTT, RCT, and strength properties of W and W alloys are determined by the deformation processes and their prehistory as well as by the doping compositions. The data base of the materials envisaged for the divertor design is affected by a large range of uncertainties and is uncompleted (especially for irradiation effects). For irradiated W the presently known temperature window range extends from 800 to 1200 °C (see also [2.1-5]). Therefore a development of W alloys to broaden this operating temperature window from the today's range to 600–1300 °C, i.e. increasing the RCT and simultaneously lowering the DBTT is required. In principle, tungsten can be alloyed with other refractory elements (e.g. Hf, Ta, Mo, Nb) and noble metals (e.g. Re, Ir, Rh). W-Re alloy, for instance, exhibits excellent DBTT and RCT behaviors in the unirradiated condition and good mechanical properties. Drawbacks in application include its strongly reduced thermal conductivity, its small resources, and its activation. The RCT of W can be improved by adding fine oxide particles (ODS tungsten), such as La_2O_3, Y_2O_3 or ThO_2. In detail, the W precursors are blended with oxides and subjected to sintering and mechanical processing to achieve high densities. The oxide dispersion strengthened (ODS) version of W by adding 1 % lanthanum oxide (WL10) is regarded the most suitable option for the divertor structures because it helps improve the RCT and machinability of pure W. It is assumed that finer grains or ODS particles will positively affect the properties, as it is known from the use of SPD (severe plastic deformation) techniques e.g. in the fabrication of very thin foils or wires. The DBTT and RCT of WL10 under fusion neutron irradiation are estimated to be around 600 and 1300 °C, respectively, being regarded the "design window" range, according to which the coolant temperature is to be adjusted. In this design helium at 10MPa and inlet/outlet temperatures of 600/700 °C is used as coolant. It is compatible with hot refractory metals and any kind of blanket systems. The use of He coolant also allows for a relatively high gas outlet temperature and, hence, a high thermal efficiency of the power conversion systems.

Supporting structures made from high temperature ODS steel: Good mechanical properties of steel for use as supporting structure are required at enough high temperature to cope with the material used in the thimble. The temperature windows should be compatible with the adjacent component material at the interface. The structural material envisaged for the DEMO reactor is ferritic-martensitic steel EUROFER. It excels by a higher strength, a higher thermal shock resistance, and a better neutron swelling behavior compared with austenitic steel 316L used in ITER. Drawbacks include the more complicated manufacture and welding and an elevated value of DBTT under irradiation which means that a minimum service temperature of the structural components is fixed at about 300 °C. The strength values and the thermal expansion coefficient of EUROFER steel are included in Table 3.8-4 as a function of the temperature. The relatively low thermal expansion coefficient and the relatively high thermal conductivity exert a favorable effect on the thermal stress behavior, which is described by the thermal stress factor σ_T [MPa.m/W] = $\alpha \cdot E/\lambda(1-v)$, with α [1/K]: the thermal expansion coefficient (TEC), E [MPa]: the Young's modulus, λ [W/mK]: the thermal conductivity, and v: the Poisson's ratio. Taking into account the creep rupture strength of EUROFER, wall temperatures must not exceed about 550 °C at the maximum. By mechanical alloying of Eurofer steel with about 0.5 wt% Y_2O_3 the ODS version of the steel is obtained, which allows higher maximum temperature limit by 100 K and thus a reasonable working temperature window (Figure 3.8-21). In the divertor design therefore ODS Eurofer (or a ferrite version of it) is used for its entire structure.

3.7 Thermal Hydraulic Design

3.7.1 Overall Thermohydraulic Layout

On the basis of an electric output of the power plant of 1500 MW, the fusion power was determined to be 3410 MW, assuming a net efficiency for the blanket cycle of 0.43 and an energy multiplication factor of 1.17 [2.1-5]. The total divertor power amounts to 583 MW. It consists of 335 MW neutron-generated heat power for the divertor bulk (256.2 MW) and the outboard (OB) and inboard (IB) target plates (44.1 MW OB, 34.7 MW IB, total 78.8 MW) and 248 MW surface heat power (alpha and heating power) for the divertor target. A power distribution between inboard and outboard targets of 1:4 given by plasma physics constraints was assumed, thus leading to a surface heat power of 49.6 MW and 198.4 MW for the inboard and outboard target, respectively (Table 3.7-1). For a 7.5° divertor cassette the size of an outboard target plate is about 810 mm x 1000 mm (toroidal x poloidal), leading to an overall average surface heat load of about 3.5 MW/m^2, i.e. 5.1 MW/m^2 for the OB target plate. These heat loads have to be managed by any divertor design.

	(A) Q_{surf} $(Q_\alpha + Q_{aux.heating})$	$Q_{neutron}$			$Q_{surf.} + Q_{neutr.}$
		Target plates	Bulk	(B) Sum	(A) + (B) Total 48 cassettes / 1 cassette
Outboard (OB)	198.4	44.1	143.5	187.6	386 / 8.042
Inboard (IB)	49.6	34.7	112.7	147.4	197 / 4.104
Sum	248	78.8	256.2	335	583 / 12.146

Table 3.7-1: Total energy balance of a model C divertor [2.1-5] in MW.

The boundary conditions for the detailed thermohydraulic divertor layout are given by a) the total heat loads and b) the position and shape of the loading curves, which depend on the strike point position (Figure 3.7-1). Its position is assumed to lie in a range between 0.1–0.5 m, measured from the bottom end of the plate. The actual power density distribution q(x) (MW/m^2) for any strike point position is described mathematically by the following equation [3.7-1, 3.7-2]:

$$q(x) \;=\; q_0 \cdot e^{-(x/a)^2} \tag{3.7-1},$$

with $q_0 = 10$ MW/m^2, x (m): distance from the strike point, a = 0.07 for x < 0, a = 0.5 for x ≥ 0.

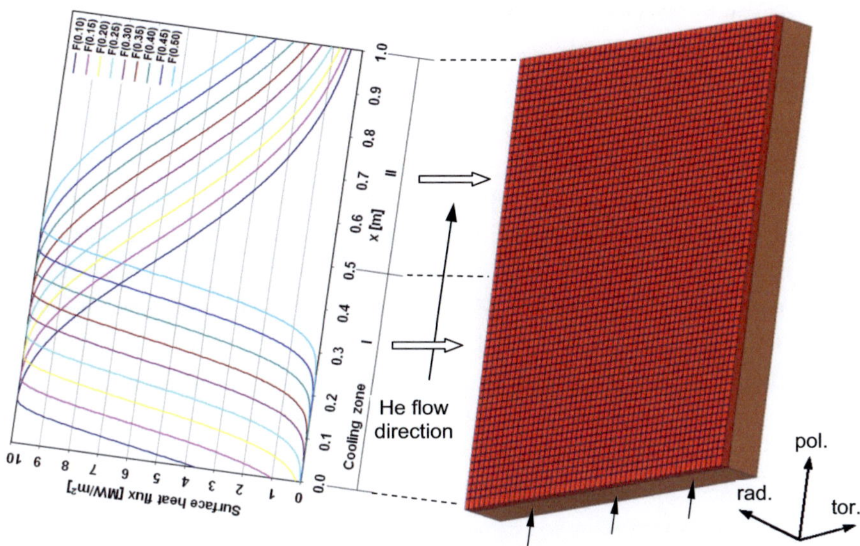

Figure 3.7-1: Poloidal surface heat load distribution assumed for the outboard target plate, x = poloidal distance (m) measured from the bottom edge of the target plate [3.7-1, 3.7-2].

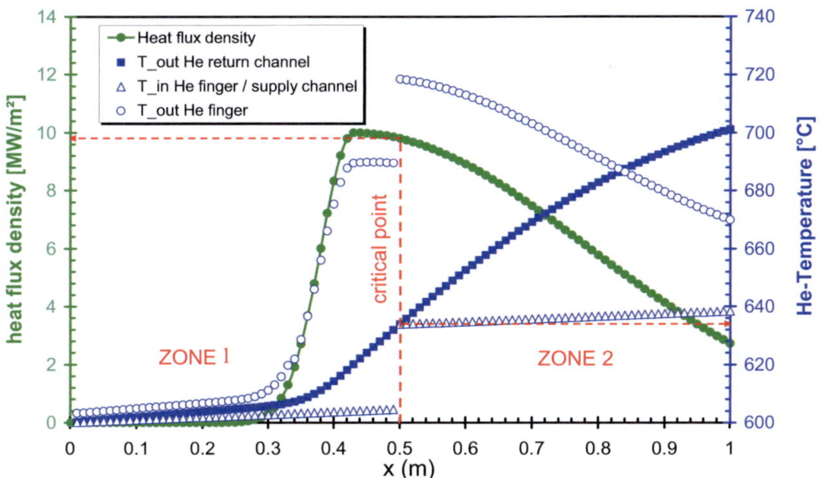

Figure 3.7-2: Coolant temperature development along the poloidal length of the target plate which is divided into two cooling zones according to Figure 3.7-1, x = poloidal distance (m) measured from the bottom edge of the target plate [3.7-1]. Image courtesy of T. Ihli.

In addition to this load, the neutronic (volumetric) load of $\approx$ 13 MW/m^3 in W and 10 MW/m^3 in steel, respectively, was taken into account. This additional heating corresponds to about 1.13 MW/m^2 on average for a 1 meter target plate length.

In contrast to preliminary studies in PPCS [2.1-5], it was decided for the plates being entered at a helium temperature of 600 °C instead of 700 °C, the lower boundary of the operation temperature window which is limited by the DBTT of tungsten. Therefore, a lower coolant temperature is possible and suitable at this stage for keeping the temperature of the thimble below the upper boundary of the temperature window defined by the RCT of tungsten alloy, which was estimated to be about 1300 °C under irradiation.

As described above, a peak load of 10 MW/m^2 at an average load of about 5 MW/m^2, i.e. a peaking factor of about 2, has been taken into account for the cooling design. A flooding of the entire plate with a twice as high mass flow rate would cause an immense pressure loss, which is proportional to the mass flow rate squared divided by the density. Therefore in this layout another solution is preferred by dividing the target plate into two poloidal cooling zones each of about 0.5 m length, which are connected in series (Figure 3.7-1). One entire outboard target plate comprises a total of 2876 1-finger modules, 1472 fingers in the cooling zone I and 1404 fingers in the cooling zone II. The cooling fingers are grouped into a larger 9-finger modular unit on a modular basis. Several 9-finger modules then form stripes, which eventually build up to a total target plate (Figure 3.6-3). The step-by-step modular design allows the separate testing of individual modules. A preliminary estimate for sufficient cooling brought a required mass flow rate of 9.6 kg/s for an outboard target plate or about 6.8 g/s per divertor cooling finger. Since the helium temperature is higher at the entrance to the zone 2 than in zone 1, a critical case can occur when the first finger of the cooling zone 2 is charged by the peak load. Therefore, in the following design calculations the worst case scenario, i.e. 6.8 g/s mass flow rate per finger and 634 °C inlet temperature, is considered. The coolant temperature development for the OB plate is shown in the Figure 3.7-2 as an example. Finally, a complete thermal hydraulics for a divertor cassette is as follows (Figure 3.7-3): The helium temperatures at the inlet and outlet of the cassette amount to 540 and 717 °C, respectively, i.e. a temperature rise of 177 K. The corresponding values for the OB plate are 600 °C and 701 °C, temperature rise 101 K, respectively. The estimated pressure loss is about 0.5 MPa pressure for the whole divertor cassette. This corresponds to a power required for the helium blower of about 9 % of the heat power to be dissipated and is still well below the defined engineering limit of 10 % (chapter 3.5).

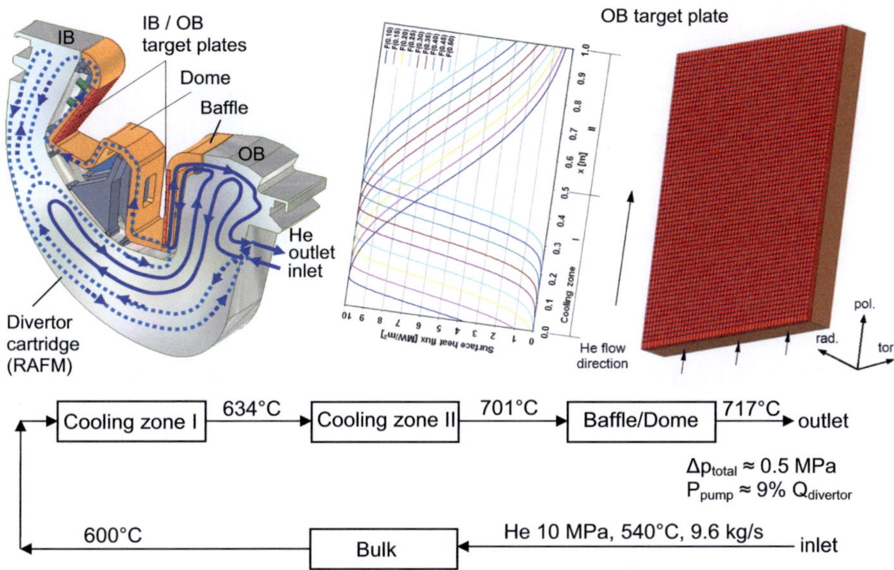

Figure 3.7-3: Overall thermohydraulics design with a schematic representation of a possible flow chart of the coolant.

3.7.2 Jet Impingement Heat Transfer

Impingement cooling is an effective way to generate a high cooling rate in many engineering applications which has now been applied to enhance heat transfer in a He-cooled divertor for a nuclear fusion power plant. The ability of controlling heat transfer from the surface by varying flow parameters, such as jet exit velocity and flow temperature, as well as geometrical parameters, such as jet exit opening, nozzle-to-surface spacing, and nozzle-to-nozzle spacing in arrays, is the key factor that has led to the sustained and widespread use of jet impingement technologies. The most commonly used geometries are axisymmetric (circular orifice or pipe) of diameter D and two-dimensional slot nozzles of width B (Figure 3.7-4).

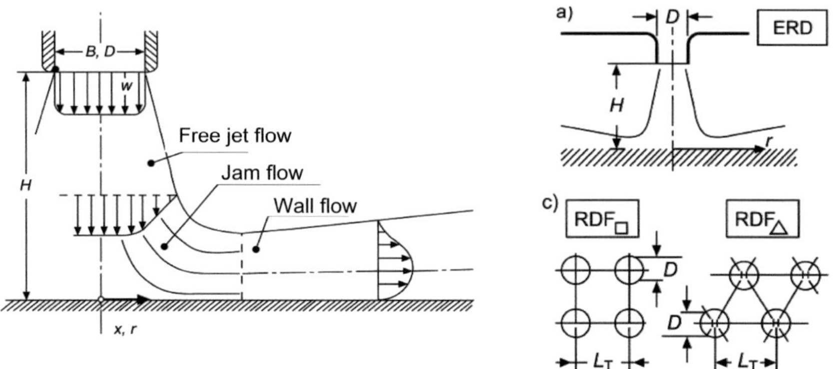

Figure 3.7-4: Flow pattern development in singlejet impingement (left) and possible arrangements for multi-jet impingement with round nozzles (right) [3.7-3].

The flow field of an impinging jet (Figure 3.7-4, left) can be divided into three zones [3.7-3]: (1) free jet flow prior to impact, (2) jam flow, and (3) wall flow. The single jet impingement cooling is sufficient for a hot spot case.

For cooling a large area, several individual nozzles can be arranged in an array of multiple jets. Figure 3.7-4 to the right shows possible arrangements of circular nozzles suitable for multi-jet impingement cooling of large areas, with the characteristic length D, the jet diameter and T, the pitch between the nozzles. The triangular arrangement (most right) corresponds to our divertor layout case, for which an estimate of the heat transfer coefficient can be derived from the following relations:

$$htc = \frac{Nu \cdot \lambda}{D} \tag{3.7-2},$$

with Nu (-): Nusselt Number, λ (W/mK): thermal conductivity of the coolant, D (m): the jet hole diameter, and for helium, which is assumed to be an ideal gas: $\lambda = 3.623 \cdot 10^{-3} \cdot T^{0.66}$ [W/mK], T in [K] [3.7-4].

The following Nusselt correlation for multi-jet impingement cooling is given in [3.7-3]:

$$Nu = G \cdot Re^{2/3} \cdot Pr^{0.42}, \tag{3.7-3}$$

with Re: the jet Reynolds number, Pr: the Prandtl number, and G: the geometry function of the form:

$$G = \frac{d^* \cdot \left(1 - 2.2 \cdot d^*\right)}{1 + 0.2\left(h^* - 6\right) \cdot d^*} \cdot \left[1 + \left(\frac{10 \cdot h^* \cdot d^*}{6}\right)^6\right]^{-0.05} \tag{3.7-4},$$

$$f = \frac{\pi}{2 \cdot \sqrt{3}} \cdot \frac{D^2}{L_T^2}\;;\; d^* = \sqrt{f} = 0.9523 \cdot \frac{D}{L_T}\;;\; h^* = \frac{H}{D} \quad \text{(see Figure 3.7-4)}.$$

Scope: $0.004 \leq f \leq 0.04$; $2 \leq h^* \leq 12$; $2000 \leq Re \leq 10^5$.

The jet Prandtl and Reynolds numbers can be calculated as follows:

$$Pr = \frac{\eta \cdot c_p}{\lambda}, \tag{3.7-5},$$

$$Re = \frac{w \cdot D_h}{\upsilon}, \tag{3.7-6},$$

with
$$\upsilon = \frac{\eta}{\rho}, \tag{3.7-7},$$

for helium assumed as an ideal gas: $\eta = 0.4646 \cdot 10^{-6} \cdot T^{0.66}$ [kg/ms], T in [K] [3.7-4],

where v (m^2/s) is the kinematic viscosity, η (kg/ms) the dynamic viscosity, λ (W/mK) thermal conductivity of the coolant, c_p (J/kg/K) the specific heat capacity = 5200 for He, and ρ (kg/m^3) the density of the coolant:

$$\rho = \frac{p}{RT}, \tag{3.7-8},$$

with p (MPa): the pressure of the coolant, R (J/kgK): gas constant = 2078.75 J/kgK for helium, T (K): temperature of the coolant.

The average jet velocity w (m/s) can be calculated from the known jet mass flow rate m_j (kg/s) and cross-sectional area of the jet holes A_j (m^2):

$$w = \frac{m_j}{\rho A_j} \tag{3.7-9}.$$

The speed of sound in helium, which is calculated from $c_{He} = \sqrt{\kappa \cdot R \cdot T}$, with $\kappa = 5/3$ for monatomic gases, amounts to about 1773 m/s at a temperature of 634 °C.

An estimate of the heat transfer performance in this multi-jet impingement cooling layout for the reference design (chapter 3.6) yields average values which are shown in Table 3.7-2.

Aj (m^2)	ρ (kg/m^3)	η (kg/ms)	ν (m^2/s)	Dj, averaged (m)	λ (W/mK)
7.6·10^{-6}	5.3	4.16·10^{-5}	7.84·10^{-6}	6.2·10^{-3}	0.3244

w (m/s)	c_{He} (m/s)	Re (-)	Pr (-)	Nu (-)	htc (W/m^{2}K)
168	1772	13350	0.67	68	35382

Table 3.7-2: Estimated flow parameters and heat transfer coefficient for the reference design with 6.8 g/s helium mass flow rate for one cooling finger at 634 °C and 10 MPa, multi-jet array geometry: 24 holes of Ø0.6 mm plus one central hole of Ø1 mm.

3.8 Computer and Experimental Simulations

3.8.1 Computational Fluid Dynamics Analysis

With the help of modern computer tools, such as combinations of the Computer Aided Design (CAD), Computational Fluid Dynamics (CFD), and Finite Element Method (FEM) applications, various design variations can be investigated on the performance and load-carrying capacity. Most practical CFD approaches rely on the solution of the Reynolds-averaged Navier–Stokes (RANS) equations. Basically, turbulence is generated in the flow field, then transported via the convection and diffusion processes, and finally dissipated by friction. To predict the turbulence especially in the near-wall region (boundary layer) by means of numerical flow simulation, a suitable turbulence model is required. The k-ε turbulence model is a widely used non-linear two-equation model, where k is the turbulence kinetic energy, a measure of the intensity of turbulence, and ε is the dissipation, i.e. consumption of turbulence kinetic energy due to friction. The standard form of the k-ε model uses a default wall function and does not require any numerical resolution of the boundary layer. This causes on the other hand some inaccuracies in depicting the velocity fields due to the assumption of isotropic turbulence. In order to adequately replicate strong anisotropy of the turbulent boundary layer, various modified turbulence models are used, which include the near-wall low turbulence region. They are classified as low Reynolds number turbulence models and multi-layer turbulence models. The latter uses multiple-layer wall functions for the outer region of the flow near the wall, the near-wall layer (i.e. the area of the logarithmic wall law), and the viscous or laminar sublayer with a linear velocity distribution.

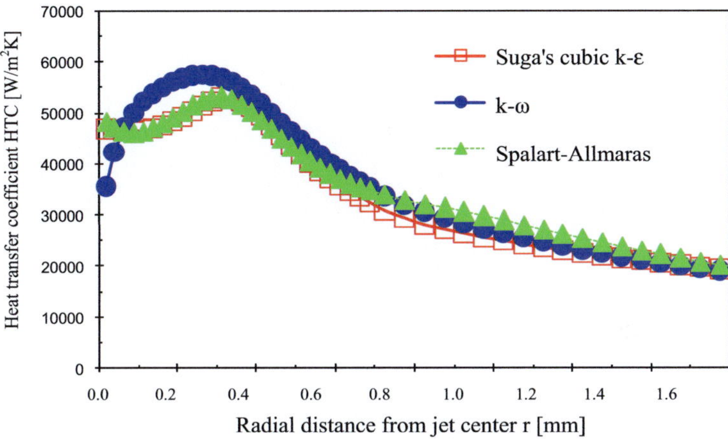

Figure 3.8-1: Example of heat transfer coefficient (low Reynolds number) for jet diameter of 0.6 mm as a function of distance r from the jet centre [3.7-5].

Figure 3.8-1 shows as a CFD computational example for a single jet impingement cooling for the present divertor thermal-hydraulic conditions and geometry, the distribution of the calculated heat transfer coefficient over the radius of single-jet under various turbulence models [3.7-5]. The considered models in this case are: the non-linear Suga's cubic k-ε model, the two-equation Wilcox k-ω model[16], and the one-equation Spalart-Allmaras model[17] (for more detailed results see [3.7-6]). While the results of the former and the latter models show good agreement regarding the location and the amplitude of the maximum of the heat transfer coefficient, the result of the k-ω model deviates significantly from the others. Only the Suga's cubic k-ε model shows, with its symmetry of the curve at the stagnation point (horizontal tangent line), the physically most meaningful result for this case. Depending on the used turbulence models in CFD analysis, the htc maximum value reaches a value of

---

[16] The k-ω model is a widely used two-equation turbulence model. Here, a transport equation for k and a transport equation for the characteristic frequency of the energy dissipating eddies are resolved.

[17] The Spalart-Allmaras model is a relatively simple one-equation model that solves a modeled transport equation for the kinematic eddy (turbulent) viscosity.

between 52,000 and 57,000 W/m^{2}K at a distance of about 0.28–0.3 mm from the jet center and decays steeply with increasing distance.

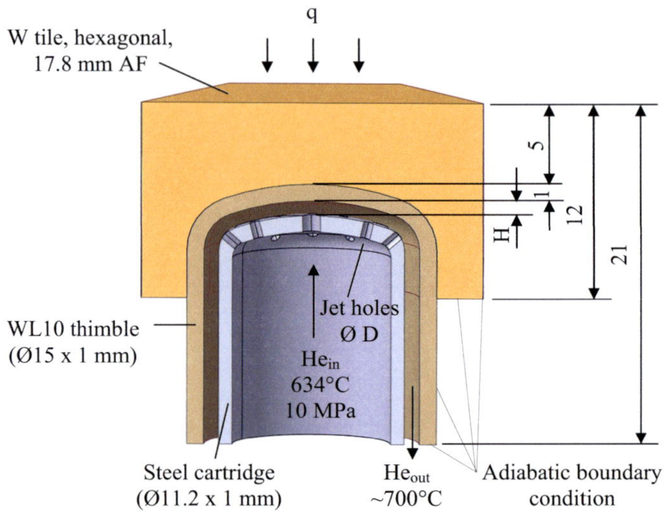

Figure 3.8-2: Geometry model for CFD and FEM calculations.

Especially for the described multi-jet impingement cooling design a comprehensive thermo-hydraulic parameter study was performed [3.8-1]. This was primarily investigated the dependence of the divertor cooling capacity on the jet coolant hole size and the jet-to-wall distance. The results of this procedure are the basis for the following design screening. Figure 3.8-2 shows the geometric model that was used for the calculation. It consists of tungsten components, a hexagonal tile (17.8 mm AF) and a thimble (Ø15 x 1 mm), in their basic form without any edge roundings. A heat load q acts uniformly on the tile top surface. The tile layer thickness between its plasma-facing surface and the connection to the thimble is 5 mm. The thimble wall thickness amounts to 1 mm. In addition, a steel jet cartridge with an array of small holes of diameter D is inserted inside the thimble. The Jet-to-wall distance is designated as H. Details and the variation of the calculation parameters D and H are compiled in the Table 3.8-1. Adiabatic boundary condition was applied to the outer unheated surfaces and the lower cut surface. The material data for the tungsten and steel solid parts are compiled in the Table 3.8-4. Boundary conditions for the helium coolant are: a rectangular profile of mass flow rate m at the inlet (temperature 634 °C and pressure 10 MPa) (He properties see chapter 3.7 above). The CFD calculations were performed using the Fluent code. The two-equation turbulence model Reynolds normalisation group (RNG) k-ε instead of the standard k-ε turbulence model was used because the latter applies only to fully developed turbulent flow at high Reynolds number. For analysis of the thimble temperature, a maximum limit on the recrystallization temperature of the WL10 material of 1300 °C was adopted.

In Table 3.8-1, the jet parameters and the calculated values of htc, temperatures and pressure losses for the nominal case (q = 10 MW/m^2, 6.8 g/s He at 10 MPa and 634 °C) are summarized. In the first variation (A–C*), the parameter jet-to-wall distance H from 0.9 to 1.2 mm is varied. The calculated maximum thimble temperature for this case as a function of H is shown in Figure 3.8-3. The result shows only a weak influence of the wall distance in the investigated range on the maximum thimble temperature which is between 1152 °C and 1164 °C. In the next category (C–E), the jet hole diameter is varied (0.4–0.85 mm), while keeping the jet-to-wall distance and the number of jet holes unchanged. As can be clearly seen from Figure 3.8-4, a change in diameter of the jet nozzle has a stronger influence on the maximum thimble temperature (1077 °C–1266 °C) than in the first case. In the last group (F–H), the number and diameter of jet holes are varied, while keeping the jet cross-sectional area constant.

Option	Jet holes: number and size D (mm)		Jet-to-wall distance H (mm)	Calculated values by Fluent			
	center	other		htc (W/m^{2}K)	$T_{max, thimble}$ (°C)	$T_{max, tile}$ (°C)	Pressure loss (MPa)
A	1 x Ø1	24 x Ø0.6	1.2	32136	1164	1711	0.135
B	1 x Ø1	24 x Ø0.6	0.6	32843	1152	1696	0.141
C	1 x Ø1	24 x Ø0.6	0.9	32422	1157	1703	0.132
C*	1 x Ø1	24 x Ø0.4	0.9	40355	1077	1635	0.538
D	1 x Ø1	24 x Ø0.7	0.9	29133	1200	1739	0.073
E	1 x Ø1	24 x Ø0.85	0.9	24776	1266	1803	0.036
F	1 x Ø1	18 x Ø0.794	0.9	28058	1210	1753	0.071
G	1 x Ø1	12 x Ø0.939	0.9	26973	1224	1773	0.073
H	1 x Ø1	6 x Ø1.212	0.9	26026	1231	1778	0.073

Table 3.8-1: Results of CFD parameter study with Fluent [3.8-1] for the reference design HEMJ [3.6-3]. Nominal load case: q = 10 MW/m^2, 6.8 g/s helium at 10 MPa and 634 °C.

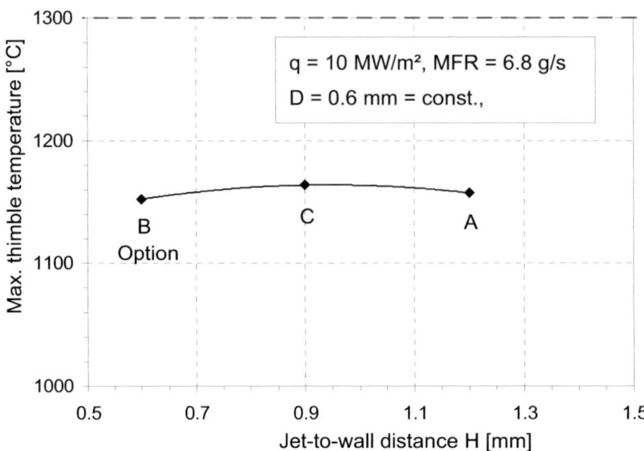

Figure 3.8-3: Max. thimble temperature [°C] as a function of jet-to-wall distance H according to Table 3.8-1. From left to right: Options B, C, and A. Courtesy of R. Kruessmann [3.8-1].

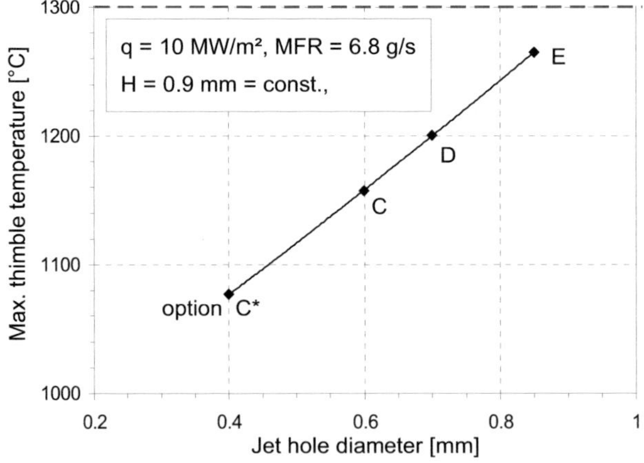

Figure 3.8-4: Max. thimble temperature [°C] as a function of jet diameter D according to Table 3.8-1. From left to right: Options C*, C, D, and E. Courtesy of R. Kruessmann [3.8-1].

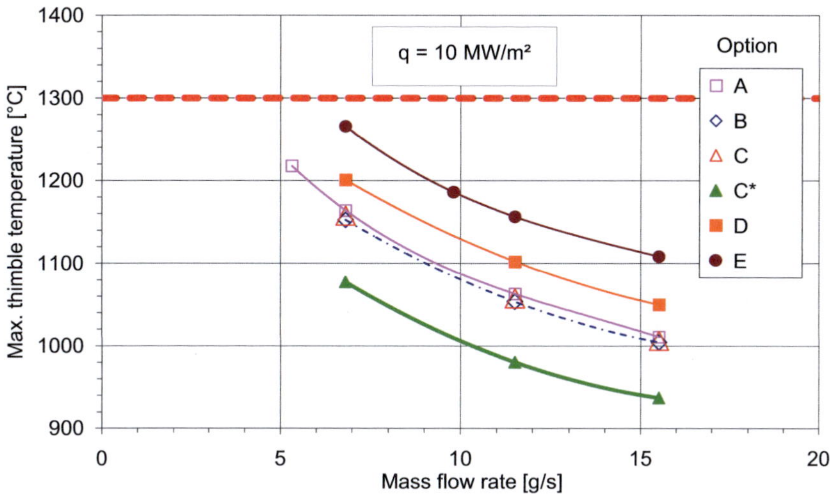

Figure 3.8-5: Max. thimble temperature [°C] as a function of mass flow rate for all the design options shown in Table 3.8-1 under a heat load of 10 MW/m². Courtesy of R. Kruessmann [3.8-1].

In Figure 3.8-5, the maximum thimble temperatures as a function of the mass flow rate at q = 10 MW/m² for all design options according to Table 3.8-1 are plotted. At a nominal mass flow rate of 6.8 g/s and under a heat load of 10 MW/m² all maximum temperatures of the thimble are well below the defined limit of 1300 °C. Except for the Option E with the largest jet diameter of 0.85 mm in this group, the remaining options have a certain reserve, which can tolerate a mass flow rate reduction down to about 5 g/s without exceeding the limit temperature.

Although a reduction of the jet cross section (option C* compared to C) leads to a lowering of the maximum thimble temperature, on the other hand it causes a strong increase in pressure loss due to the quadratic dependence of the pressure loss on flow velocity. All in all, the C option is a good compromise between the maximum structure temperature (Figures 3.8-4, 3.8-5) and the pressure loss (Figure 3.8-6). It is therefore chosen as a reference for further testing. The calculated heat transfer coefficient of about 32000 W/m²K (Figure 3.8-7) agrees well with the assessment in chapter 3.7.2. Furthermore, a comparative calculation for the reference case with a different code ANSYS/CFX [3.8-2] brought a good agreement (Figure 3.8-8). The results yield a maximum thimble temperature of about 1170 °C at a maximum jet velocity of about 240 m/s resulting in about 0.12 MPa pressure losses. The jet velocity reached corresponds to about 0.13 Ma and is at the bottom of the subsonic range (<0.75 Ma). The He velocity in the jet hole can be effectively reduced by slight enlargement of the drill hole diameter of a few percent.

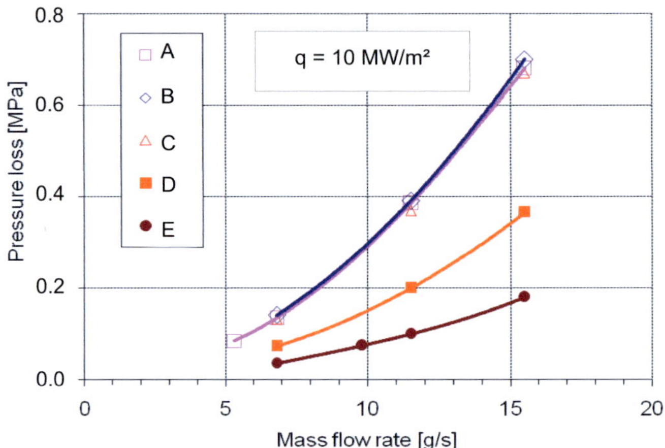

Figure 3.8-6: Pressure loss [MPa] as a function of mass flow rate according to Table 3.8-1. Options A through E. Courtesy of R. Kruessmann [3.8-1].

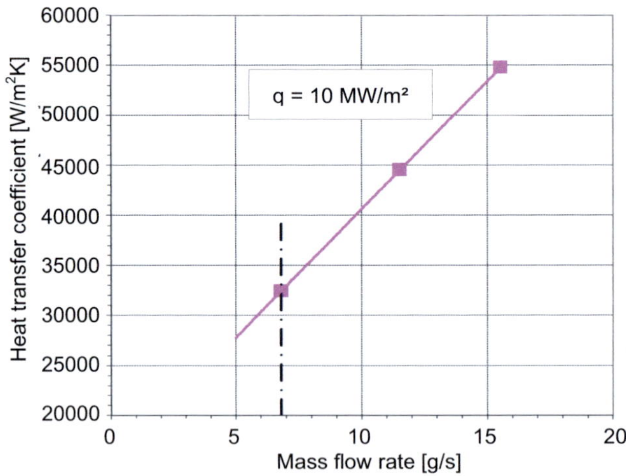

Figure 3.8-7: Heat transfer coefficient [W/m^{2}K] of the option C as a function of mass flow rate according to Table 3.8-1. E. Courtesy of R. Kruessmann [3.8-1].

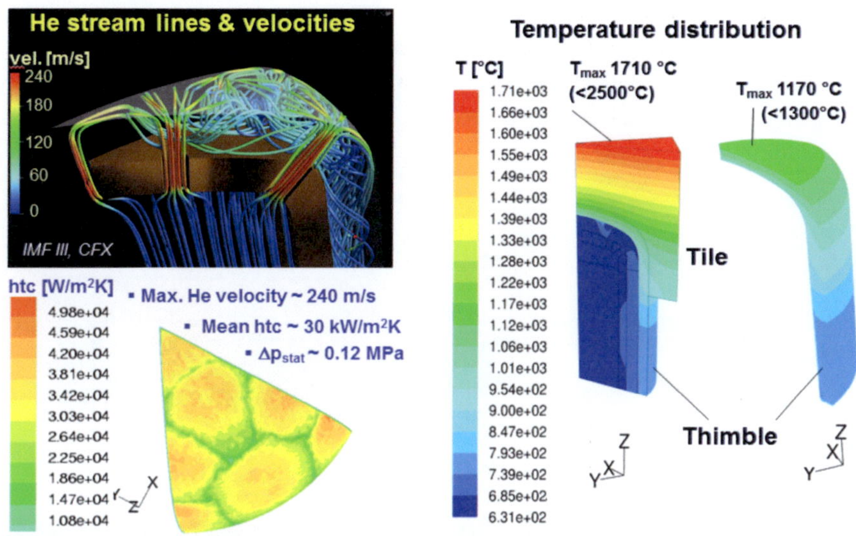

Figure 3.8-8: Results of CFD simulations with ANSYS/CFX for the reference design (option C). Image courtesy of R. Kruessmann [3.8-1].

3.8.2 Experimental Verification of the CFD Simulations

In collaboration with Georgia Tech (GT), Atlanta, USA, an instrumented HEMJ mock-up has been designed for CFD code validation testing in the air loop at GT and the helium loop HEBLO at KIT. Figure 3.8-9 shows the construction of such a mock-up [3.8-4]. It consists of three main parts: a) the test element including a jet cartridge and a concentric thimble, b) the tee, which provides the fittings for the coolant and the instrumentation to monitor the flow, and c) the copper body with an integrated electrical heater. The bottleneck shape of the copper block ensures a compact and uniform axial heat flux at the thimble surface. The thimble and cartridge were both made from C36000 free-machining brass, which has a similar thermal conductivity to tungsten alloy. Two identical mock-ups were manufactured by GT and mounted at these different air and helium test loops (Figure 3.8-10). The corresponding test conditions for the experiments in helium and air circulation are shown in Table 3.8-2, while maintaining the same Reynolds number as in the DEMO reference case. Thermocouple (TC) probes (Ø0.5 mm, type E in GT air loop and type K (Ni-Cr/Ni) in

HEBLO) were used. Four of them are inserted into the brass thimble at varying depths spaced by 90° to measure the temperature distribution over the cooled surface. The main thermocouple positions TC1 to TC4 in the brass thimble, and TC5 to TC7 (copper block) are illustrated in Figure 3.8-11.

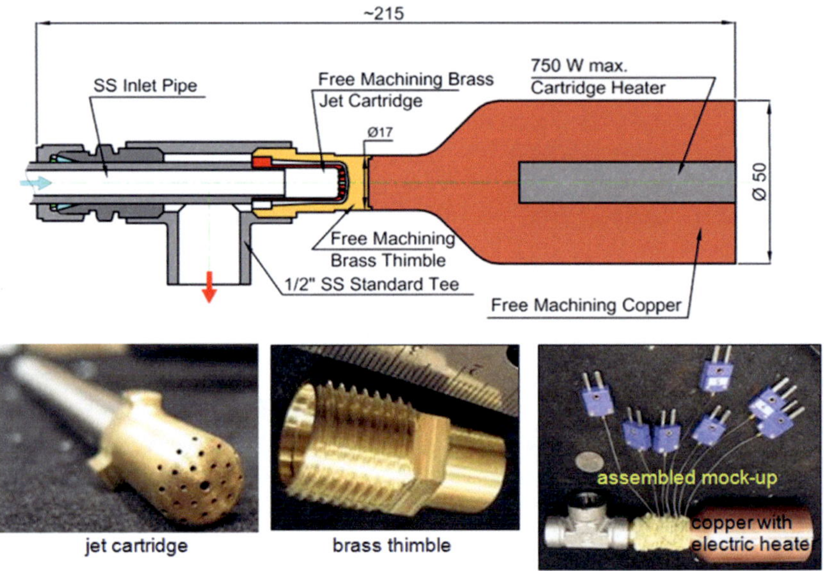

Figure 3.8-9: 1:1 Mock-up design for CFD code validation (dimensions in mm).

	Coolant	T_{in} (°C)	P_{in} (MPa)	η 10^{-5} (kg/ms)	Heat flux (MW/m^2)	Mass flow rate (g/s)	Re (–)
DEMO	Helium	634	10	4.16	10	6.8	21400
HEBLO	Helium	35	8	2.04	2	3.33	21400
GT	Air	20	0.724	1.85	1	3.03	21400

Table 3.8-2: Test conditions in accordance with the nominal DEMO case, assuming the same Reynolds number.

Figure 3.8-10: Integrated mock-up in test facilities: a) air loop at GT (left), b) helium loop HEBLO at KIT (right).

Position	measured	calculated with ANSYS/CFX
TC1 (°C)	120	125
TC2 (°C)	134	135
TC3 (°C)	153	142
TC4 (°C)	137	141
TC5 (°C)	196	195
TC6 (°C)	212	209
TC7 (°C)	225	224
TC8 (°C)	288	282
TC9 (°C)	288	282
T_{in} (°C)	38	38
T_{out} (°C)	51	50
ΔT (K)	13	12
P_{in} (MPa)	7.85	7.78
P_{out} (MPa)	7.76	7.76
ΔP (MPa)	0.018	0.015

Table 3.8-3: Comparison between numerical and experimental HEBLO results [3.8-5]. Test parameters: Q_{Heater} = 227 W, mfr = 3.6 g/s, q = 1 MW/m^2. TC position according to Figure 3.8-11.

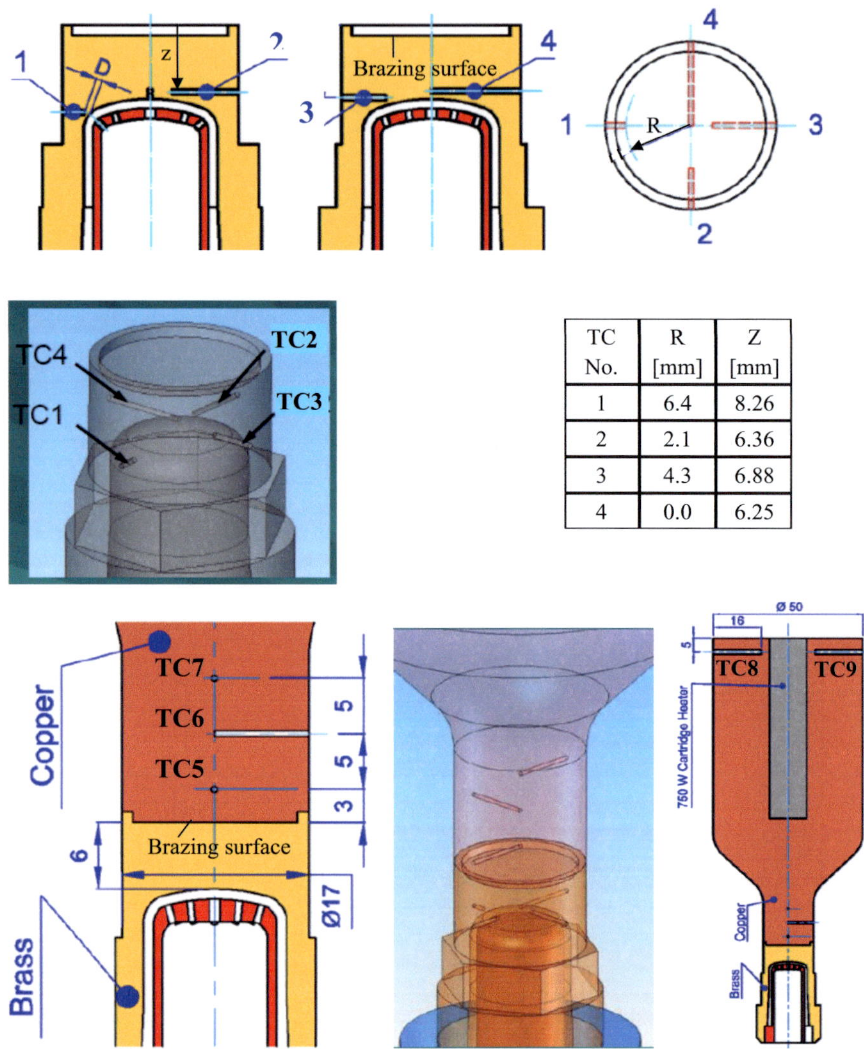

TC No.	R [mm]	Z [mm]
1	6.4	8.26
2	2.1	6.36
3	4.3	6.88
4	0.0	6.25

Figure 3.8-11: Position of the thermocouples TC1 to TC4 (brass thimble) and TC5 to TC9 (copper block). R: radial distance from the centre (mm); z: axial distance from the brazing surface (mm): D: thermocouple diameter = 0.5 mm.

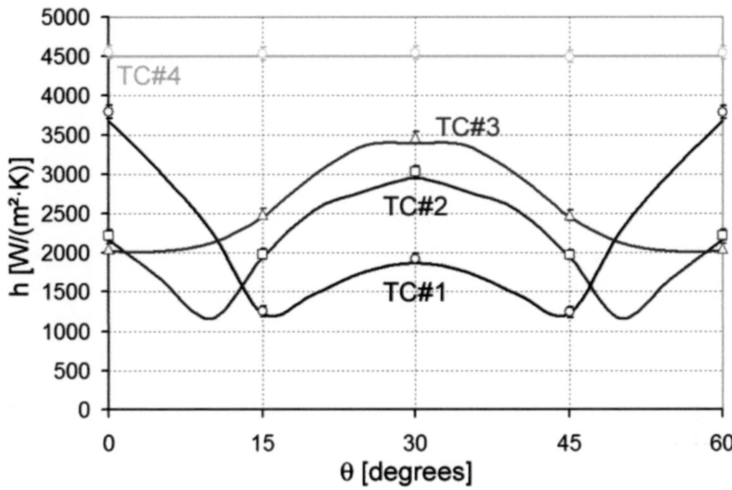

Figure 3.8-12: Azimuthal surface HTC profiles from the experimental measurements in air loop (symbols) and predictions (lines) at Re= 21,400 [3.8-4].

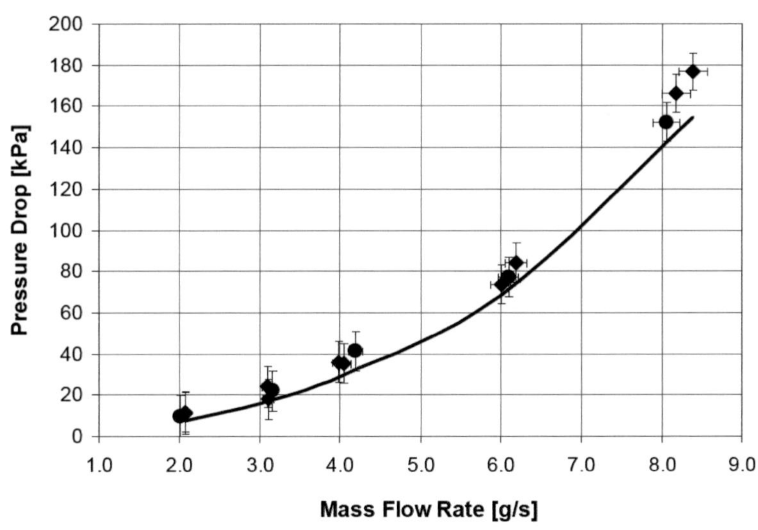

Figure 3.8-13: Experimental and numerically predicted pressure loss across the test section (air loop). Solid line: FLUENT® prediction; experimental results for 182 W (diamond symbol) and 227 W (round symbol) power input [3.8-5].

61

For all test conditions, good agreement was obtained in air loop between the model predictions and the experimental data. Differences between the calculated and measured temperatures are small (< 2 K), so that the corresponding effect on the surface HTC is nearly negligible (Figure 3.8-12); in all cases, the experimental and predicted HTC differ by less than 5 %. In the helium experiments [3.8-5] differences between the temperature calculated with CFX and the measured values amount up to about 11 K (Table 3.8-3). The calculated pressure loss of CFX was underestimated by about 17 % compared to measurement, while the calculated results with FLUENT show a less underestimation of about 12 % (Figure 3.8-13).

3.8.3 Structural Analysis with Finite Element Method (FEM)

After the thermal conditions (i.e. temperature distribution) in the divertor cooling fingers are known, the resulting stresses in such engineered and designed structural components must now be examined and optimized if necessary in the next step. For the reference design, detailed stress analysis using the program ANSYS/Workbench [3.8-6] with subsequent design optimization for stress-reduction was carried out in [3.8-7]. The evaluation of the existing stresses was performed according to the ASME Code [3.8-8] in the same manner as in the DEMO blanket design studies [3.8-9] as follows:

$$\sigma_{eq} < \sigma_{adm} \tag{3.8-1},$$

where σ_{eq}, σ_{adm} are the equivalent stress and the admissible stress, respectively,

σ_{adm} = Sm,t for primary membrane stresses, average takenover the cross section,

σ_{adm} = 1.5 Sm,t for primary membrane plus bending stresses,

σ_{adm} = 3 Sm,t for primary and secondary stresses,

with Sm = min (2/3·Rp0.2, 1/3·Rm) and Sm,t = min (Sm, 2/3·Ru,t, Rp1.0,t).

The symbols have the following meanings: Rp0.2 is the offset yield strength, Rm is the ultimate tensile strength, Ru,t is the creep rupture strength, Rp1.0,t is the 1 % proof stress, and t is the operating lifetime, with $t \approx 1.75 \cdot 10^4$ h assumed for the divertor by about 2 years.

The equivalent stress intensity σ_{eq} used in the comparison with the admissible stress (σ_{adm}) is the von Mises stress derived according to the hypothesis of shape change:

$$\sigma_{eq} \;=\; \sqrt{\frac{(\sigma_1 - \sigma_2)^2 + (\sigma_2 - \sigma_3)^2 + (\sigma_3 - \sigma_1)^2}{2}} \tag{3.8-2},$$

where σ_1, σ_2, σ_3 are the principal stresses. The primary and secondary stresses are superposed at the stress component level.

In Table 3.8-4, the material data for tungsten materials and ODS steel, respectively, are shown as a function of temperature. For simplicity the Sm values of T91 steel (as a substitute Eurofer) (see also [2.1-5]) at an elevated temperature level of 100 K were adopted for ODS steel. For tungsten material, the creep strength as well as radiation-induced material properties, the latter for both tungsten and steel, are initially not taken into account due to the unavailability of data. Some thermo-physical and mechanical properties such as thermal conductivity and yield strength of tungsten and WL10 are also shown graphically in Figures 3.8-14 and 3.8-15.

For stress analysis, the same geometric model as for CFD calculations in Figure 3.8-2 is used. In addition, a 'frictionless support' boundary condition is applied to the lower cutting plane, which means that this plane remains plane and parallel. After transfering the heat transfer coefficient from ANSYS/CFX into ANSYS/Workbench, a temperature calculation was first conducted by ANSYS/Workbench itself. The calculated temperature distribution together with the helium internal pressure then act as loadings on the finger structure in the following stress calculation. Based on the initial geometry, Figure 3.8-16 shows the results of the temperature (left) and stress calculations (right). The maximum temperatures calculated with ANSYS/Workbench amount to 1754 °C in the tungsten tile and 1187 °C in WL10 thimble, respectively. They are slightly higher than the results calculated with ANSYS/CFX (Figure 3.8-8), probably due to the use of average heat transfer coefficients. The resulting maximum stress in the thimble is approximately 369 MPa occuring in the round corner on its inner side at a temperature of ~968 °C, which is below the allowable 3-Sm limit of ~378 MPa. For the tungsten tile the maximum stress is about 345 MPa. It occurs locally at the lateral vertical tile edges at a prevailing temperature of about 1149 °C and is well below the allowable 3-Sm limit of about 465 MPa. Even an existing stress of about 158 MPa on the tile surface at an extremely high temperature of about 1724 °C is still below the allowable 3-Sm value of about 169 MPa.

Properties	@ T [°C]	W	WL10	@ T [°C]	ODS EUROFER
λ [W/mK]	20	173	123	20	25.9
	500	133	107	200	28.1
	1000	110	97	400	29.2
	1500	101	94	600	28.5
ρ 10^3 [kg/m3]	20	19.3	19.3	20	7730
	500	19.2	19.2	200	7680
	1000	19.0	19.0	400	7610
	1500	18.9	18.9	600	7540
cp [J/kgK]	20	129	126	20	449
	500	144	146	200	523
	1000	158	153	400	610
	1500	170		600	755
ρ_{el} 10^{-6} [Ω.m]	20	0.05		20	0.50
	500	0.18		200	0.95
	1000	0.32		400	1.06
	1500	0.49		600	1.16
E [GPa]	20	398		20	206
	500	390		200	194
	1000	368		400	182
	1500	333		600	151
v [-]	20	0.28		20	0.3
	500	0.28		200	0.3
	1000	0.29		400	0.3
	1500	0.30		600	0.3
TEC 10^{-6} [1/K]	20	4.0	4.6	20	10.4
	500	4.2	4.8	200	11.2
	1000	4.5	5.0	400	11.9
	1500	4.8	5.1	600	12.5
Rp0.2,min [MPa]	20	1360		20	400
	500	854	430	500	338
	1000	465	362	600	293
	1500	204	197	700	204
Rm,min [MPa]	20	1432	854	20	580
	500	966	538	500	471
	1000	565	373	600	395
	1500	266	201	700	273
Sm [MPa]	20	477	284	20	193
	500	322	179	500	174
	1000	188	124	600	146
	1500	89	67	700	101

Table 3.8-4: Thermophysical and mechanical properties of W, WL10 [3.8-10], and ODS steel [2.1-5].

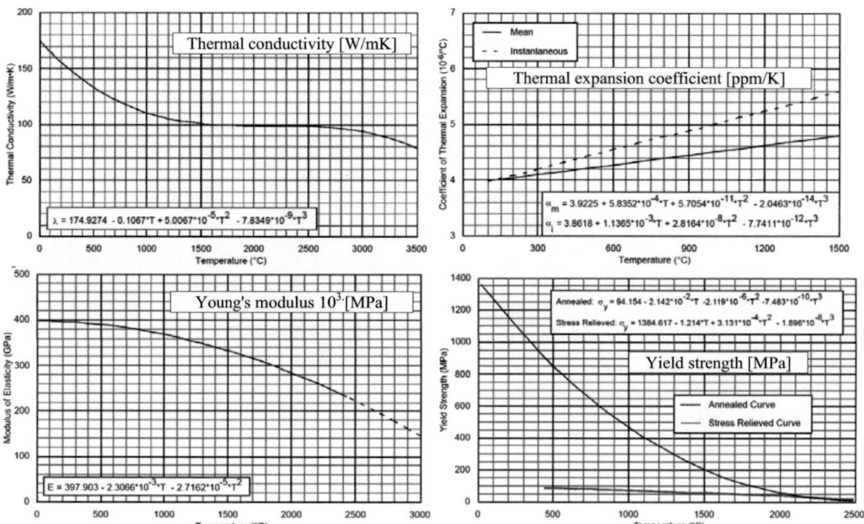

Figure 3.8-14: Some thermophysical and mechanical properties of tungsten as a function of temperature T in [°C], according to ITER MPH [3.8-10].

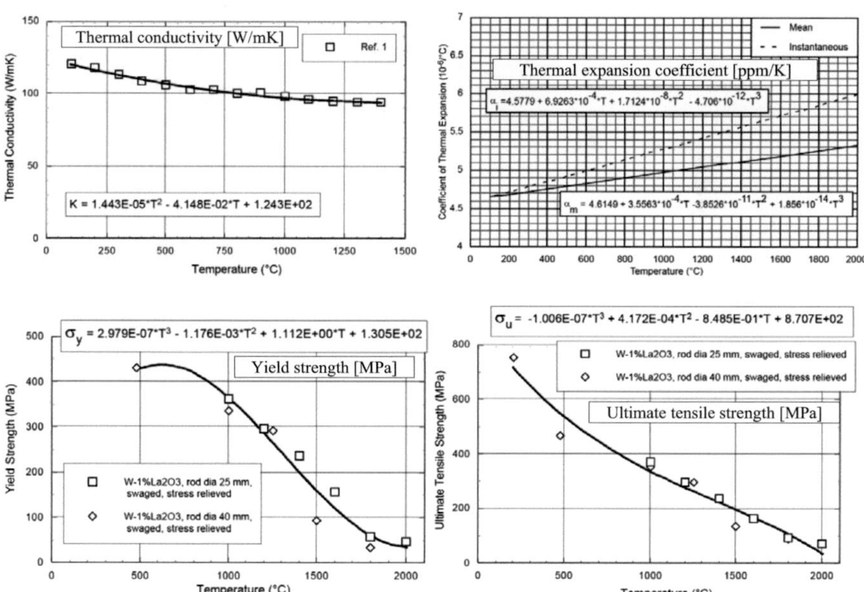

Figure 3.8-15: Some thermophysical and mechanical properties of WL10 as a function of temperature T in [°C], according to ITER MPH [3.8-10].

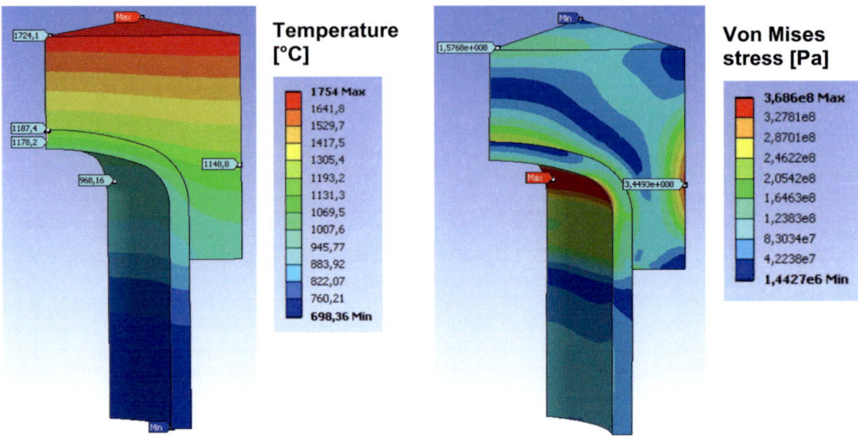

Figure 3.8-16: Calculated temperature and stress distributions with ANSYS/Workbench for the basic geometric model according to Figure 3.8-2 [3.8-7].

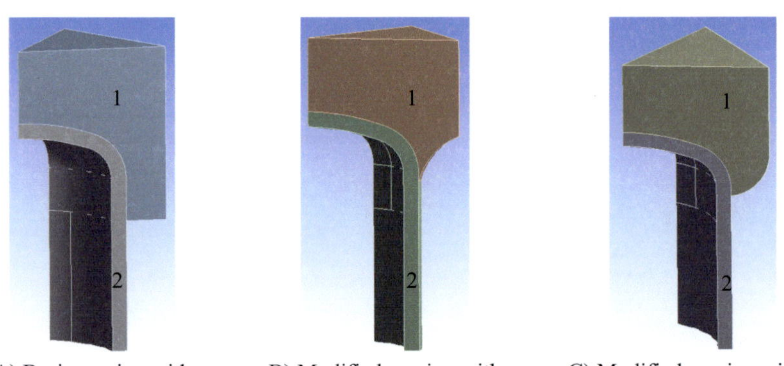

A) Basic version with straight tile shoulder

B) Modified version with concave chamfered tile shoulder

C) Modified version with a convex rounded tile shoulder

Figure 3.8-17: Some variations of the outer contour of tile (B and C) on the basis of the initial geometry A [3.8-7]. (1: W tile, 2: WL10 thimble).

In the course of iterative improvements to the design based on the experimental findings (see later point), the basic geometry was further optimized in order to reduce thermal stresses. This was done using the parameter variations, where a number of parameters were investigated (for details see [3.8-7]). One of the main parameters is the tile outer contour, which contributes significantly to thermal stress reduction. Figure 3.8-17 shows as an example the most effective modifications of the tile outer contour (B and C) derived from the basic form (A), which brought very good results.

Figure 3.8-18 shows the temperature and stress results for case B with chamfered tile shoulder. Here, a more uniform stress distribution throughout the components by the modified tile geometry is clearly visible. The maximum stress in the tile was reduced by more than 50 % to about 230 MPa and in the thimble by about 7 % to ~345 MPa, respectively. The maximum component temperatures have risen slightly ($T_{max,\ tile}$ ~1806 °C, $T_{max,\ thimble}$ ~1202 °C) due to a decrease of the lateral heat-conducting area. However, they remain still well below the allowable limit.

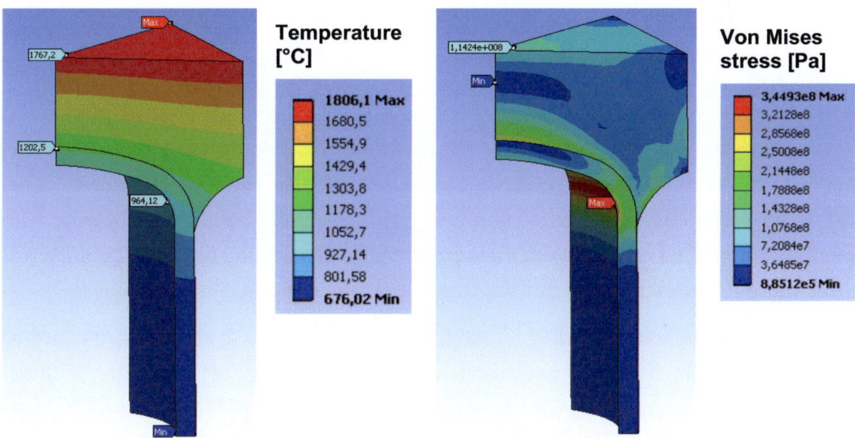

Figure 3.8-18: Calculated temperature and von Mises stress distributions with ANSYS/Workbench for the modified version with concave chamfered tile shoulder [3.8-7].

A similar trend of improvement is shown in Figure 3.8-19 for case C with a convex rounded tile shoulder. Here in this case, the maximum temperatures ($T_{max,\ tile}$ ~1782 °C, $T_{max,\ thimble}$ ~1200 °C) increased relative to the base model less than in case B due to its higher degree of filling in the shoulder area of the tile. The decrease in the maximum stresses in the tile remains about the same as in the case B, while in the thimble, a reduction of the maximum stresses of about 10 % is determined. However, a relatively high peak stress of up to 371 MPa appears in the lower tile edge at contact with the thimble. For this reason, the B version is preferred.

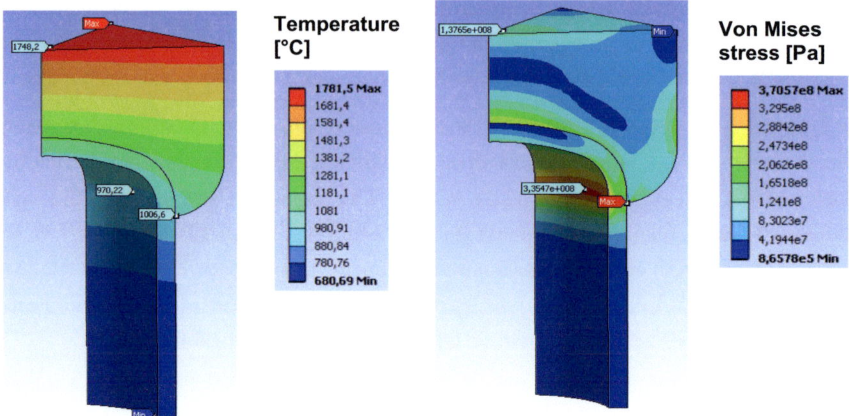

Figure 3.8-19: Calculated temperature and von Mises stress distributions with ANSYS/Workbench for the modified version with a convex rounded tile shoulder [3.8-7].

Another parameterization option is the castellation of the tile surface (i.e. the type, number, width and depth of segmentation). The best results of the individual parameter variations were merged and integrated into a new geometry. Together with the above-described results of the main parameter variation, the optimum geometry of the cooling finger obtained in this way can collectively be described as follows:

- A hexagonal tile with a width across flats of 17.8 mm with chamfered shoulder, having a total sacrificial layer thickness of 5 mm and a three-section castellation with a groove width and depth of 0.5 mm and 2.75 mm, respectively;

- A thimble with an outer diameter of 15 mm and 1 mm wall thickness, having a dished boiler head.

Figure 3.8-20 shows the temperature distribution calculated by ANSYS for the optimized design under nominal conditions. The maximum temperatures of 1175 °C in the thimble and 1720 °C in the tile, respectively, agree well with the values from the CFD calculations (Figure 3.8-8). The calculated equivalent von Mises stress for this case amounts to 280 MPa (location: at the bottom region of the tile at the connection area to the thimble) which is lower than the value of the original HEMJ model C of 369 MPa. Both values lie below the allowable stress limit of 373 MPa (at elevated temperature of 1300 °C) according to ASME [3.8-8]. It can be seen that safety margin to the permissible stress has been increased. Figure 3.8-21 shows the load-oriented reference designs considering the operating temperature window of the divertor. The resulting temperatures and stresses have been checked and are below the permissible limits.

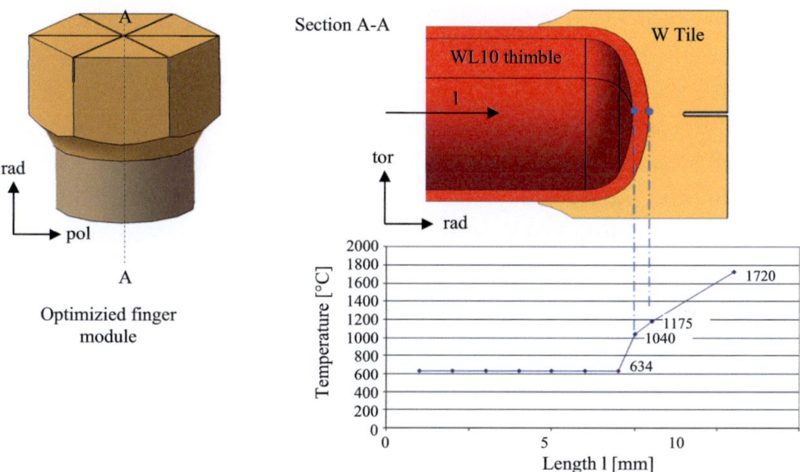

Figure 3.8-20: Temperature distribution in the optimized cooling finger calculated with ANSYS/Workbench. Boundary conditions: q = 10 MW/m^2, 6.8 g/s He at 10 MPa and 634 °C.

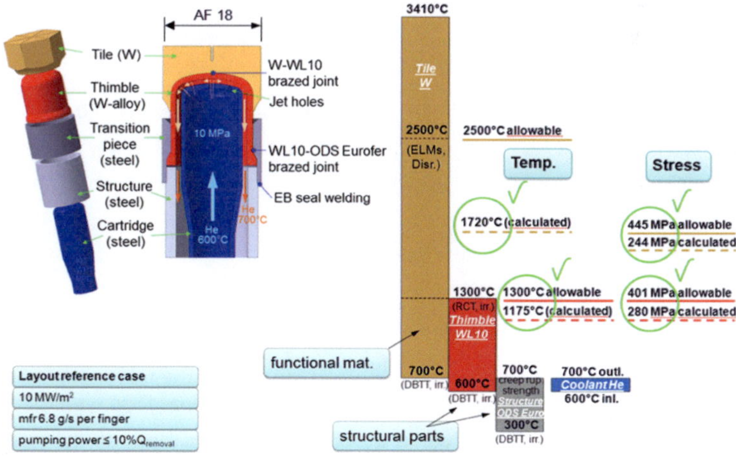

Figure 3.8-21: The HEMJ reference design layout and working temperature window.

3.9   Experimental Verification of the Design

3.9.1   Combined Helium Loop and Electron Beam Test Facility at Efremov

In collaboration with Efremov Institute St. Petersburg, Russia, a combined helium loop and e-beam high heat flux (HHF) test facility was built there during 2004–2005 with a goal of experimental proof of principle. This facility enables mock-up testing under DEMO relevant conditions, i.e. a surface heat load of at least 10 MW/m^2 and corresponding helium cooling conditions with inlet temperature of 500 °C, inlet pressure of 10 MPa and mass flow rate in the range of ~5–15 g/s. The flow chart diagram of the helium loop is shown in the Figure 3.9-1. It is a closed loop which consists of a main circuit with circulating helium and an auxiliary circuit for filling and evacuating the loop. Figure 3.9-2 illustrates the main internal components of the loop, the EB vacuum vessel, and the mock-up holder device. The mock-up is fixed to a holder and connected to the helium loop. A water-cooled copper mask with a central hexagonal hole (span size 18mm) is mounted concentrically around the mock-up in order to protect the mock-up holder structure made from steel and the thermocouples at the W-steel connection from e-beam damage. An additional TZM cage is placed between the mock-up and the Cu mask, which serves to absorb the excess electron scanning power at the edge. Following diagnostics were used: Measuring of absorbed heat load by ΔT calorimeter, measuring of incident heat load by spot calorimeter, temperature measuring by infrared (IR) camera, thermocouples for calibrations and bulk temperature measurements, and video camera.

The EB facility had until 2008 an older gun with a power of 60 kW and an acceleration voltage of 27 kV and was fitted with a new gun EH 200V (Figure 3.9-3, bottom right) with a capacity of 200 kW and an acceleration voltage of 40 kV. Electrons are accelerated in the gun electrostatic field to 30 % of the light velocity (at 40 kV accelerating voltage). In the vacuum chamber, the electron beam interacts with the atoms of the surface of the target test object. Most of the kinetic energy of the EB is converted to thermal energy in a thin layer of micron thickness, which is used to simulate the surface heat load on the test object.

Figure 3.9-3 (top) shows the complete assembly of the helium loop with the EB test facility. The helium loop is installed on a vehicle that is movable on the rails. This allows flexibility and rapid change of the experiments. Since the position of the mock-up is at the edges of the vacuum vessel, the electron beam is projected sideways at an angle of about 10 degrees to the surface normal of the sample via a mirror system (Figure 3.9-3, a). The desired power density is pre-set using a water-cooled copper calorimeter. The actual power absorbed by the mock-up is determined by the temperature rise of helium. Furthermore, the facility has an automatic device to simulate the hot shutdown and startup operations by cyclic switching on and off the EB gun (Figure 3.9-3, b). The cycle length was chosen (e.g. 30s beam on/30s beam off) so that the maximum temperature in the test mockup is reached in the quasi steady state. The calibration of the infrared camera was done in this case using tungsten calibration sample equiped with thermocouples. Calibration plots are obtained by heating the sample by an electon beam.

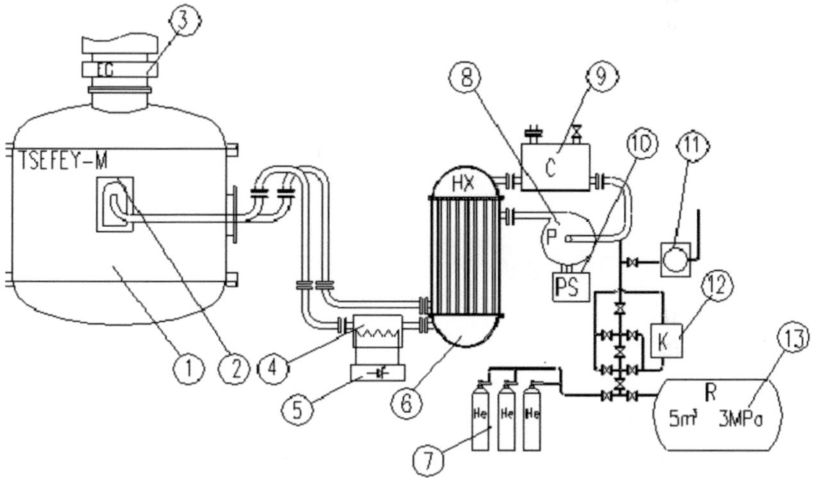

Figure 3.9-1: Closed helium loop scheme: Efremov.1 Vacuum chamber, 2 DEMO mock-up, 3 Electron beam gun, 4 Resistance heater, 5 Heater power supply, 6 He Recuperator/heat exchanger (HEX), 7 He ballooncylinders, 8 He blower, 9 Cold water HEX, 10 Blower power supply, 11 Loop evacuation pump, 12 Compressor for loop filling and evacuating, 13 He tank – receiver. Not included: valves, oil traps, loop diagnostics, loop control, external water cooling, industrial power supply, etc.

Figure 3.9-2: Built-in components of the helium loop and the TSEFEY vacuum tank. Control panel: 1 – filter, 2 – throttling orifice for gas flow measuring, 3 – gas pressure regulator (output pressure 10 MPa), 4 – motorized valve, 5 – manual valve, 6 –gas flow regulator, 7 – safety valve, 8 – heater line, 9 – cooler line, 10 – measuring differential pressure, 11 – to feeding balloon cylinder bank, 12 – to vacuum line, 13 – to receiving balloon cylinder bank.

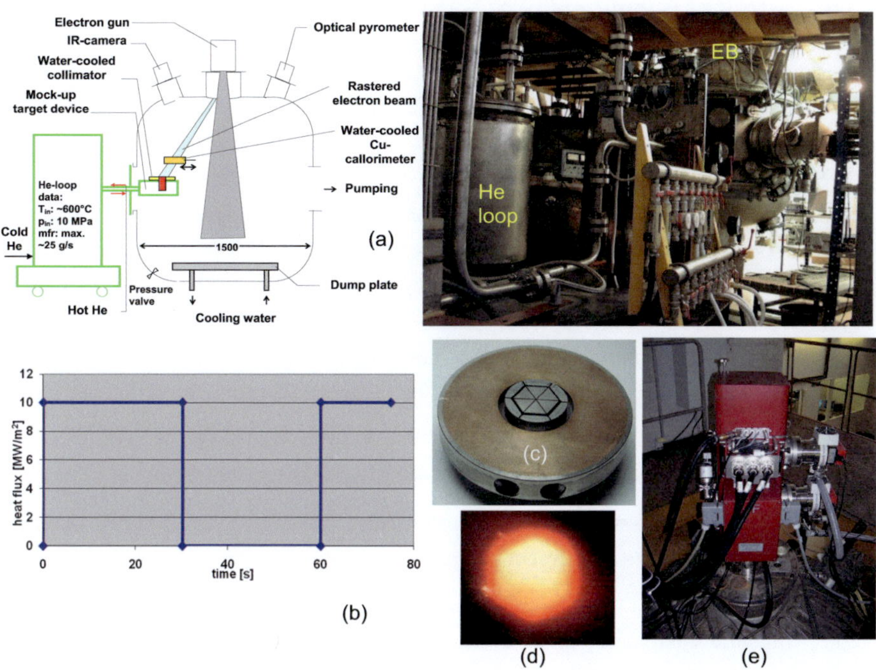

Figure 3.9-3: Top: Scheme (a) and general view of the combined He loop & E-beam test facility at Efremov. Bottom: Heat load cycle (b); Mock-up with holder (c); IR image at 10 MW/m^2 (d); New installed EB gun 2009 (EH 200V, 200 kW, 40 kV) (e).

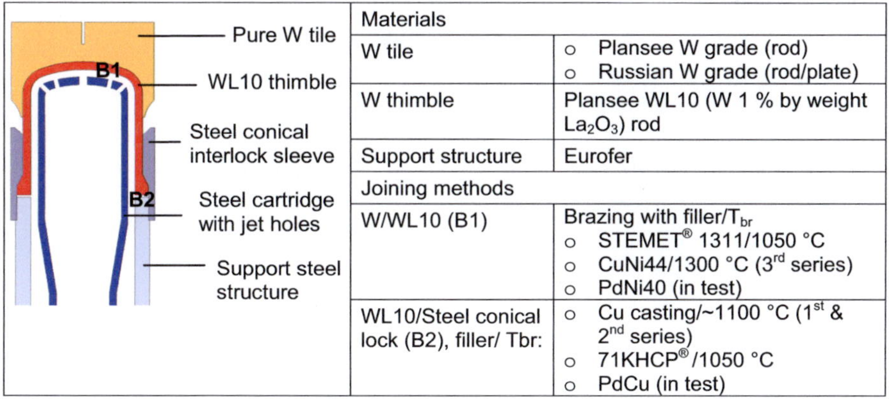

Materials		
W tile	o Plansee W grade (rod) o Russian W grade (rod/plate)	
W thimble	Plansee WL10 (W 1 % by weight La$_2$O$_3$) rod	
Support structure	Eurofer	
Joining methods		
W/WL10 (B1)	Brazing with filler/T_{br} o STEMET$^®$ 1311/1050 °C o CuNi44/1300 °C (3rd series) o PdNi40 (in test)	
WL10/Steel conical lock (B2), filler/ Tbr:	o Cu casting/~1100 °C (1st & 2nd series) o 71KHCP$^®$/1050 °C o PdCu (in test)	

Figure 3.9-4: Definition of 1-finger mock-ups for high heat flux testing.

3.9.2 Proof of Functionality of Basic Design

Before conducting the high heat flux tests, technological studies on mock-up manufacturing [3.9-1] were first performed in cooperation with Efremov. The focus of this study was in particular next to the machining of tungsten parts the joining of divertor components. Figure 3.9-4 shows the mock-up definition containing the variations of the brazing materials for the joints as well as different tungsten material grades. As shown in Figure 3.9-5, first successful attempts in joining of W parts with curved surface (top) using Ni-based filler metal STEMET® 1311 and W-steel parts with copper filler (bottom). Both joints passed preliminary tests with thermo-mechanical cyclic loading and additional internal pressure of 10 MPa in the latter case well. The functionality of these joints must be confirmed through the following actual high heat flux tests. Thus, the following motivations for the first test series arise:

o   Proof of principle and performance of the basic design (heat removal capacity).

o   Check of the thermal-hydraulic performance (pressure losses, temperatures) of 1-finger modules at different cooling regimes.

o   Investigation of thermal-mechanical behavior of 1-finger divertor elements (materials and joints) under cyclic heat load (hot shutdown simulation).

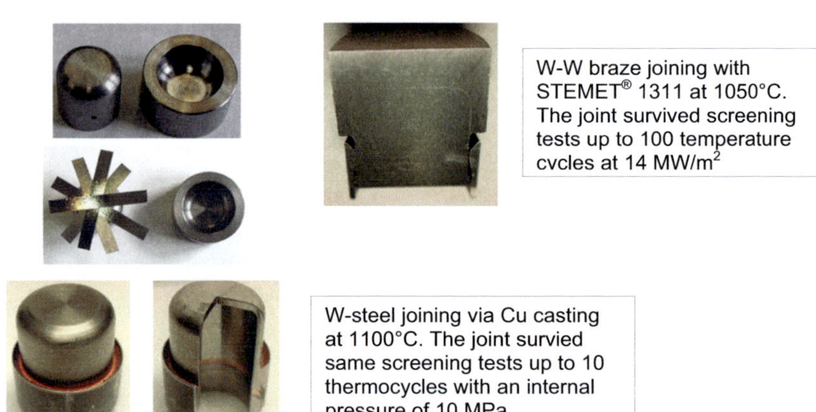

Figure 3.9-5: First successful attempts in joining of W parts with curvature (top) using a Ni-based filler metal STEMET® 1311 and W-steel parts with copper filler (bottom). Courtesy of Efremov.

Figure 3.9-6 shows the fabricated mock-ups for the first test series [3.9-2] which was performed in 2006. This test campaign contained six mock-ups (five of HEMJ and one of an old design with slot type, so-called HEMS). Castellated and non-castellated tiles made from PM tungsten rods (mock-up #1 of Plansee grade, #2-6 of Russian grade) were investigated. Thimble is exclusively made from Plansee WL10 rod material. The jet cartridge and the holding structure of the mockups are made from Eurofer. For the brazing of W-WL10 joint and WL10-steel joint with conical lock STEMET® 1311 (Ni-based) and 71KHCP® (cobalt-based) filler materials, respectively, were used. For comparison purpose, copper filling was applied for the WL10-steel transition joint in a mockup. The test campaign for the investigated mockups (sorted by the experimental order) is listed in Table 3.9-1. The mock-ups were tested within a HHF range of 5–13 MW/m^2. The temperature cyclic loading was simulated by means of switching periodically the beam on and off (variations 30s/60s, 30s/30s, and 60s/60s). The helium cooling parameters are 10 MPa inlet pressure, ~500–600 °C inlet temperature and the mass flow rate (mfr) varied in a range of ~5–15 g/s. The experimental execution and results are described below in further detail and the evaluation of the results summarized at the end.

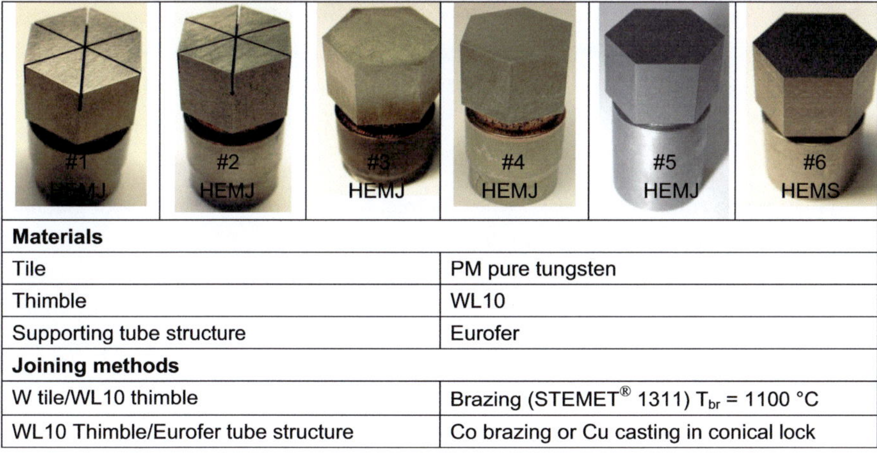

Materials	
Tile	PM pure tungsten
Thimble	WL10
Supporting tube structure	Eurofer
Joining methods	
W tile/WL10 thimble	Brazing (STEMET® 1311) T$_{br}$ = 1100 °C
WL10 Thimble/Eurofer tube structure	Co brazing or Cu casting in conical lock

Figure 3.9-6: First HHF test series 2006: Mock-up details.

Mock-up #1 (castellated): Screening tests were performed with a step-by-step increase of the applied heat flux from 5 to 9.7 MW/m^2 at a constant mfr of ~9 g/s. An early gas leakage was detected, starting after 5 TM cycles at q = 7.9 MW/m^2 ($T_{max, tile}$ ~1340 °C measured at $T_{in, He}$ ~600 °C, Figure 3.9-7, a). The tests were continued up to 10 cycles under the same heat load. They were terminated after 2 cycles at 9.7 MW/m^2 ($T_{max, tile}$ ~1550 °C and Δp ~0.17 MPa measured at $T_{in, He}$ ~600 °C). After the last shot, crack at the top and the side of the tile with penetration of brazing alloy through tile surface without gas leakage (Figure 3.9-7, b) and gas leakage through the thimble near the steel ring (Figure 3.9-7, c) were detected. Good results: no recrystallization and no cracks were found in tile and thimble area and in the conical WL10-steel joint (Figure 3.9-7, d-e), the helium loop was intact.

Mock-up #2 with a castellated W tile outstandingly survived up to 11.5 MW/m² at mfr ~13.5 g/s; $T_{in, He}$ ~550 °C. The tile temperature measured was ~1600 °C (Figure 3.9-8), the pressure loss ~0.38 MPa (Table 3.9-1). This pressure loss is equivalent to about 0.08 MPa at the nominal mass flow rate of 6.8 g/s and regarded optimistic compared to the value calculated. At a higher load of 12.5 MW/m^2, cracks in the tile and thimble and gas leakage were detected. The gas leakage was found to come from the top area of tile slots, the cracks occurred at the tile side and penetrated the thimble wall between the tile and the steel ring. Crack propagation in the thimble came from the inside. Two tile castellation segments were molten when gas leakage occurred, probably because of the increased heat flux density caused by beam focusing in the last shot. Gas leakage through the side of the tile indicates that the crack can propagate through the brazing layer. The WL10-steel joint and the He Loop remained intact.

Mock-up #3 (non-castellated): Cyclic thermal loading was performed with gradual increase in the applied heat flux ~5–9 MW/m^2. The flow rate was kept constant at 7 g/s, corresponding to the nominal value of the DEMO design. The mock-up survived 10 thermal cycles at 5.2 and 6.5 MW/m^2, but only a few cycles at q = 9 MW/m^2. Then, a crack at the tile side (Figure 3.9-9, b) and gas leakage through the tile/thimble interface (Figure 3.9-9, c) were detected. A maximum tile temperature of 1530 °C (Figure 3.9-9, a) and a pressure loss of 0.1 MPa (Table 3.9-1) were measured at 7 g/s mfr and $T_{in, He}$ of about 590 °C. Good results: no recrystallization and no cracks were found in tile and thimble area and in the conical WL10-steel joint (Figure 3.9-9, d-e), the helium loop was intact.

Mock-up #4 with a non-castellated W tile survived stepwise heat loads from 4 up to 11 MW/m^2 (mfr ~13.5 g/s, $T_{in,\ He}$ ~540 °C) each with 10 temperature cycles (60s/60s) without any damage. A maximum tile surface temperature of ~1600 °C was measured at q = 11.6 MW/m^2 using an IR camera (Figure 3.9-10, a), which agrees well with the prediction in [3.8-1] of about 1750 °C for the nominal case (10 MW/m^2, 6.8 g/s mfr, and $T_{in,\ He}$ ~634 °C). A pressure loss of 0.32 MPa was measured at the experimental mass flow rate (Table 3.9-1), which is equivalent to a value of 0.085 MPa for the nominal case after an extrapolation. It is slightly less than the measured value of 0.10 MPa in a gas puffing (GPF experiment [3.8-3] and is significantly smaller than the predicted values by the CFD codes Fluent [3.8-1] and Flotran [3.8-3] of about 0.13–0.14 MPa. The mock-up was then subjected to higher heat loads to find out its maximum performance. After six cycles at the last power step of 13 MW/m^2, an overheating of the mock-up surface was observed, which led to tile surface melting (Figure 3.9-10, b). Tile and thimble were detached due to overheating in part (Figure 3.9-10, c), confirming the relatively long cooling time of the mock-up. Cracks on each tile flank as well as small cracks in thimble growing from inside were observed (Figure 3.9-10, d-e). Absolutely no gas leak was detected, which means that the pressure-carrying thimble, the WL10 thimble-steel connection, and the helium loop remain intact.

Mock-up #5 (non-castellated) was the one whose WL10-steel joint was brazed with Co-alloy, which is much harder than copper. It survived 100 cycles (15 s/15 s) at 9 MW/m^2 and mfr ~13.5 g/s without damage. A surface tile temperature of ~1490 °C was measured (Figure 3.9-11, a), whereas the calculated value was ~1400 °C. The measured pressure loss was 0.29 MPa (Table 3.9-1). Then, the mass flow rate was reduced to ~7 g/s. The mockup resisted additional 24 load cycles at the same power until a tile surface melt (Figure 3.9-11, b) and gas leak were detected. In a subsequent post-examination cracks in the thimble (Figure 3.9-11, c) and coarse grains in the tile as a result of tungsten melt (Figure 3.9-11, d) were found.

Mock-up #6 (HEMS, Co-brazed thimble/steel joint) was tested with a gradual increase of the applied heat flux from 4.5 to 9.5 MW/m^2 (60 s/ 60 s) at a constant mfr of ~10 g/s. No visible damage and gas leak were detected. Then, the mock-up survived 100 load cycles performed with a shorter load frequency (30 s beam on and 30 s beam off) at the same mfr as well as another 100 cycles of the same kind, but at an mfr reduced to ~8 g/s. After a further decrease of the mfr down to 6 g/s at a lowered $T_{in, He}$ of ~500 °C, the mock-up survived 70 thermal cycles at a power level of 8–10 MW/m^2, giving a total number of 300 cycles applied to this mock-up. A pressure loss of 0.5 MPa was measured at an mfr of 10 g/s (Table 3.9-1), which is about a factor of three larger than the values of the HEMJ mock-ups. A maximum tile temperature of about 1720 °C was measured at $T_{in, He}$ of about 600 °C (Figure 3.9-12, a). After the last shot, cracks were detected at the top and side of the tile together with a gas leak and penetration of brazing alloy at the surface (Figure 3.9-12, b and c).

Summary of the first series of experiments:

1. Already the results of the first test series have confirmed the above first objective. The required divertor performance of 10 MW/m^2 can be achieved by He jet cooling.

2. Neither sudden destruction, completely broken mock-ups (brittle failure) nor recrystallization of W thimble was observed in any mock-up.

3. However, the results of destructive post-examinations also revealed some critical points relating to high thermal stresses and inadequate manufacturing quality, which crucially affect the lifetime of the divertor cooling finger. The latter in detail, several tungsten mock-up parts including the thimble contained pre-existing defect, presumably micro cracks [3.9-5] initiated during the fabrication processes.

4. The measured pressure losses for the HEMJ design (mock-ups #1-5) were regarded optimistic compared to the calculated value (~50 % overestimation). It was found a higher pressure loss in the slot design HEMS compared to HEMJ by a factor of approximately 3. Therefore, this option is not taken into closer consideration.

Measures to be taken:

o Optimizing the design to counteract the high thermal stresses.

o Surface treatment after manufacture for minimization of the microcracks.

Mock-up no./type[a]	Cycle number at heat flux(MW/m^2)/(beam on/off, default 30/60 s)	mfr variation (g/s)	T$_{He}$ in/out (°C)	Δp (MPa) at mfr
#4/HEMJ[b]	10 at each 4, 6, 10, and 11; 6 at 13 (→ tile detached, no leakage)	13.5	520–570/550–600	0.32
#2/HEMJ[b]	10 at each 6, 10, 11.5, and 12.5; 35 (→ cracks in tile & thimble, He leakage)	13.5	520–580/540–630	0.38
#5/HEMJ[c]	10 at each 3.8, 6, 7.3, and 9, (i); 100 at 9/(15/15), (i); 24 at 9/(15/15), (ii) (→ cracks in tile & thimble, He leakage)	(i) 13.5, (ii) 7.0	580–605/600–635	(i) 0.29, (ii) 0.08
#3/HEMJ[b]	10 at 5.2; 10 at 6.5; 5 at 9; (→ cracks in tile & thimble, He leakage)	7.0	570–580/610–640	0.10
#1/HEMJ[d]	10 at each 4.8, 6.2, and 7.9; 2 at 9.7 (→ cracks in tile & thimble, He leakage)	9.0	600/650	0.17
#6/HEMS[e]	10 at each 4.5, 5.7, 7.8, and 9,5; 100 at 9/(30/30), (iii); 100 at 9/(30/30), (iv) (→cracks at the top and side of the tile, gas leak and penetration of brazing alloy at the surface)	(iii) 10.0, (iv) 7.5	600/650	(iii) 0.50, (iv) 0.3

[a] W tile, W-WL10 joint, WL10-Eurofer joint.
[b] Non-castellated, STEMET® 1311 brazing, Cu casting in conical lock.
[c] Non-castellated, STEMET® 1311 brazing, Co brazing in conical lock.
[d] Castellated, STEMET® 1311 brazing, Cu casting in conical lock.
[e] Castellated, STEMET® 1311 brazing, Co brazing in conical lock.

Table 3.9-1: First HHF 1-finger test series 2006 performed (mockup details see Figure 3.9-6): Test campaign, test conditions used, and short results.

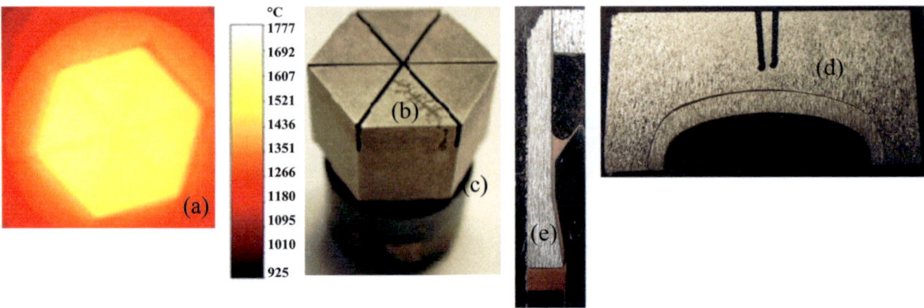

Figure 3.9-7: First HHF test series, mock-up #1 (HEMJ, castellated). (a): IR temperature image at ~9.7 MW/m^2, mfr ~9 g/s; $T_{in, He}$ ~600 °C (T_{max} ~1550 °C).

Results:

(i) Mockup survived 10 cycles each at 5,6,8 MW/m^2 and 1 cycle at ~9.7 MW/m^2 w/o damage;

(ii) Gas leakage through tile/thimble interface after 2 cycles at 9.7 MW/m^2;

(iii) Failures detected are: crack at the top and the side of the tile with penetration of brazing alloy through tile surface w/o gas leakage (b), cracks in the thimble between tile and steel ring with gas leakage (c).

Good points: No visible cracks and recrystallization in tile and thimble area (d), He loop and WL10-steel joint are intact, no cracks in conical lock area (e).

Figure 3.9-8: First HHF test series, mock-up #2 (HEMJ, castellated). (a): IR temperature image at 11.5 MW/m^2, mfr ~13.5 g/s; $T_{in, He}$ ~550 °C (T_{max} ~1600 °C).

Results:

(i) Mockup survived 10 cycles each at 6, 10, 11.5 MW/m^2 w/o damage;

(ii) Gas leak appeared at screening step with ~13.5 MW/m^2;

(iii) Failures detected are: melted W tile surface (b), cracks in tile and thimble (c, d), brazing alloy penetrated at the side surface.

Good points: He loop and WL10-steel joint remained intact.

Figure 3.9-9: First HHF test series, mock-up #3 (HEMJ, non-castellated). (a): IR temperature image at ~9 MW/m^2, constant mfr ~7 g/s; $T_{in, He}$ ~590 °C (T_{max} ~1530 °C).

Results:

(i) Mockup survived 10 cycles at 5.2 MW/m^2, 10 cycles at 6.5 MW/m^2, and a few cycles at 9 MW/m^2 w/o damage;

(ii) Gas leakage through tile/thimble interface after 5 cycles at 9 MW/m^2;

(iii) Failures detected are: crack at tile side (b), cracks in tile and thimble (c).

Good points: No cracks and no recrystallization in tile and thimble area (d), He loop and WL10-steel joint remained intact, no cracks in conical lock area (e).

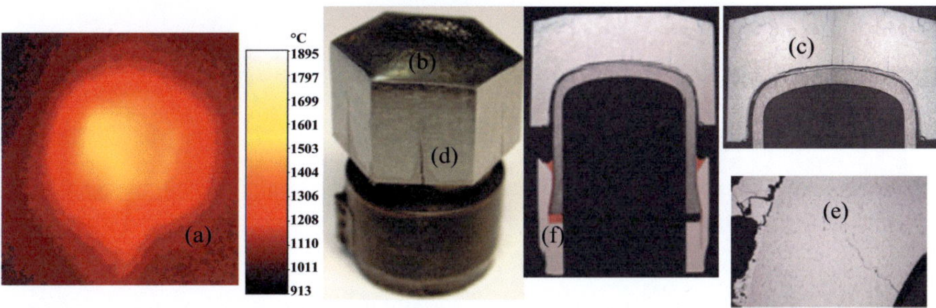

Figure 3.9-10: First HHF test series, mock-up #4 (HEMJ, non-castellated). (a): IR temperature image at 11.6 MW/m^2, mfr ~13.5 g/s, $T_{in,He}$ ~540 °C, T_{max} ~1600 °C.

Results:

(i) Mockup survived 10 cycles each at 4, 6, 10, 11 MW/m^2 w/o damage;

(ii) Failure in tile and W/W joint after 6 cycles at ~13 MW/m^2 detected: melted W tile surface (b), tile partially detached from thimble (c), cracks on each tile side (d), small cracks in thimble growing from inside (e).

Good points: No any leaks, which means that WL10-steel joint (f), thimble, and He loop remained intact.

Figure 3.9-11: First HHF test series, mock-up #5 (HEMJ, non-castellated). (a): IR temperature image at ~9 MW/m^2, mfr ~13.5 g/s; T$_{in, He}$ ~600 °C (T$_{max}$ ~1490 °C).

Results:

(i) Mockup survived 100 cycles at 9 MW/m^2, 13.5 g/s mfr w/o damage;

(ii) Gas leakage through thimble after additional 24 cycles at 9 MW/m^2 and reduced mfr of 7 g/s; (iii) Failures detected are: melted W tile surface (b), cracks in thimble (c), coarse grains in the tile (d).

Good points: He loop and WL10-steel joint remained intact.

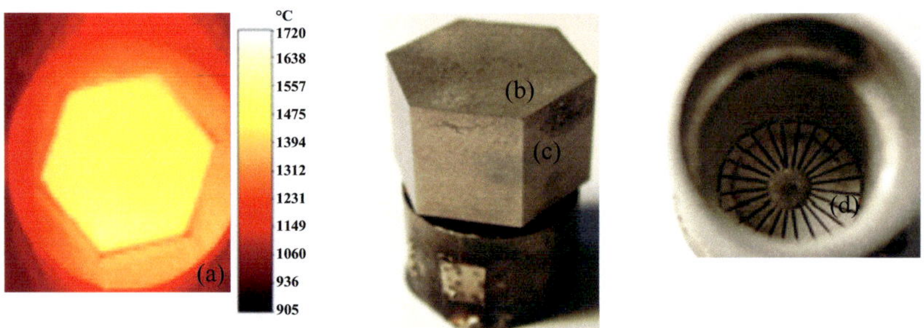

Figure 3.9-12: First HHF test series, mock-up #6 (HEMS design with slot flow promoter (d), non-castellated). (a): IR temperature image at ~9 MW/m^2, mfr ~10 g/s; T$_{in, He}$ ~600 °C (T$_{max}$ ~1800 °C).

Results:

(i) Mockup survived 200 cycles at 9 MW/m^2;

(ii) Then, failures are detected: micro-cracks area on the tile surface (b) and visible cracks on the tile sides (c), gas leak through the tile at several spots on the top surface and sides;

(iii) Relatively high pressure drop (0.5 MPa @ 10 g/s mfr).

3.9.3 High Heat Flux Testing of Advanced Design

For the following second test series 2007 [3.9-3] technological/technical improvements were made: a) the mock-up geometry was optimized to reduce the thermal stresses by means of finite element analyses (see chapter 3.8-3), b) new target device for 1-finger mock-ups was designed and manufactured which allows for changing the mock-ups without cutting and rewelding, and c) additional grinding process was applied after turning the W mock-up parts.

Ten HEMJ mock-ups (#11–20) manufactured for the second test series in 2007 are illustrated in Figure 3.9-13, whereby the mock-ups #11 and 16 were only used for the metallographic analysis without HHF tests. Testing conditions and the 2007 HHF tests results are summarised in Table 3.9-2. The mock-ups were tested at 10 MPa He, 550 °C inlet temperature, mfr = 9–13 g/s, with thermal cycling at 10 MW/m^2. A beam on/off sharp ramp of 30/30s was applied to all mock-ups for simulating the thermal cyclic loading, with the exception of the last test with mock-up #18, performed with a soft ramp (20 s – up, 20 s – hold, 20 s – down, 20 s – pause). Here, the influence of the type of power transients on the load of mock-up structure shall be investigated.

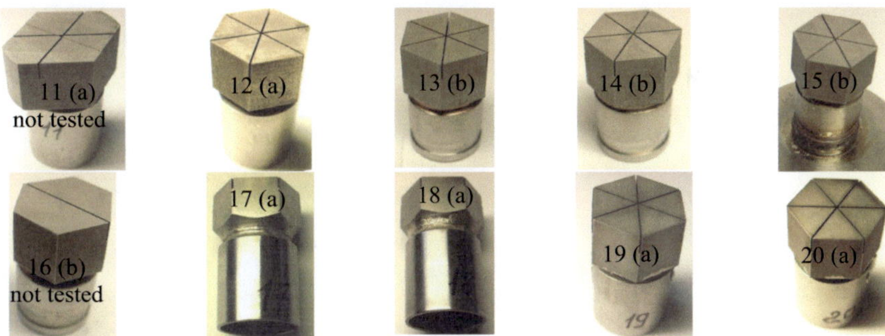

Figure 3.9-13: Second HHF test series 2007: Mock-up details.

Tile material: Plansee rod, vertical grain orientation; regular machining (turning and grinding) of tile and thimble; Thimble/conic sleeve joining: (a) Co brazing (71KHCP®, 1050 °C), (b) Cu casting (1100 °C). Beam on-off cycles: *default 30/30 s, **soft ramp: 20s–up, 20s–hold, 20s–down, 20s–pause.

Mock-up #12 with a castellated W tile was tested at a constant mfr of ~9–10 g/s. It survived 18 cycles at 10 MW/m^2. A maximum tile temperature of 1750 °C (Figure 3.9-14, a) and a pressure loss of 0.2 MPa (Table 3.9-2) were measured at 9 g/s mfr and $T_{in, He}$ of about 560 °C. Gas leak appeared at the central area of the loaded tile surface with a slightly melted spot (Figure 3.9-14, b). No remarkable visible damages were detected.

Mock-up #13 with a castellated W tile survived up to 70 cycles at 9 MW/m² at mfr ~9 g/s, $T_{He,in}$ ~570 °C. Slight cracking of the tile top surface with gas leak was detected (Figure 3.9-15).

Mock-up #14 (castellated) withstood 90 cycles at 9 MW/m². Its surface temperature increased during cycling. Finally, the tungsten tile detached from the thimble leading to further overheating and melting of the W tile. The experiment was stopped. In the **mock-ups #15, 19 and 20**, gas leak was detected during the screening tests. They were thus not testable.

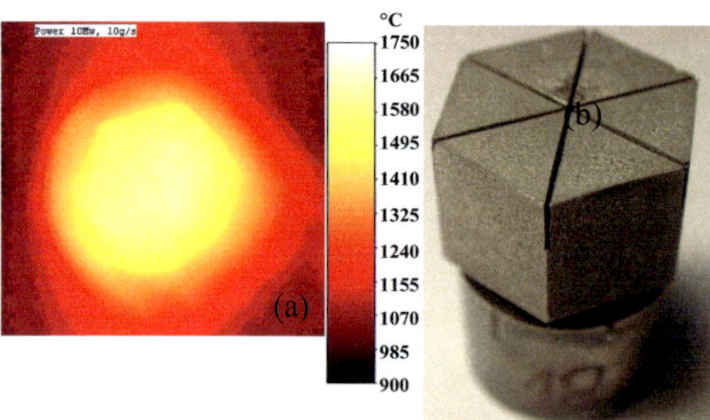

Figure 3.9-14: Second HHF test series, mock-up #12 (HEMJ, castellated). (a): IR temperature image at 10 MW/m², mfr ~9.5 g/s; $T_{in, He}$ ~560 °C (T_{max} ~1750 °C).

Results:

(i) Mockup survived 18 cycles at 10 MW/m² w/o damage;

(ii) Gas leak occurred in the central area of the tile surface with a slightly melted spot (b).

Good points: He loop and WL10-steel joint remained intact.

Figure 3.9-15: Second HHF test series, mock-up #13 (HEMJ, castellated).

Results:

(i) Mockup survived 70 cycles at ~9 MW/m², mfr 9 g/s; $T_{in, He}$ ~570 °C (T_{max} ~1600 °C);

(ii) Slight cracking of the tile top surface with gas leak at the central area of the tile were detected).

Good points: He loop and WL10-steel joint remained intact.

Mock-up #17 with the optimized tile geometry [3.8-7] was successfully tested at 89 cycles at a heat flux of 10 MW/m^2. The experiment was terminated after detecting tile temperature increase. No gas leakage and no damage occurred. The mock-up was perfectly intact (Figure 3.9-16, b). The measured pressure losses at 9 g/s mfr stayed in a range of about 0.16–0.18 MPa (Table 3.9-2) which agreed well with the values obtained from the first test series. The tile surface temperatures of this mock-up during the tests interpreted from infrared pictures reached at a range between 1600 and 1700 °C (Figure 3.9-16, a).

Mock-up #18 with the same optimized geometry as the mockup #17 was subjected to the same heat load of 10 MW/m^2 but at an increased mfr of 12.5 g/s in order not to exceed the remelting temperature of the W–WL10 brazing layer of about 1340 °C. In addition, a soft ramp which was regarded more realistic to the DEMO condition was applied in this test at the same time. This mock-up outstandingly withstood 102 thermal cycles without any damages (Figure 3.9-17).

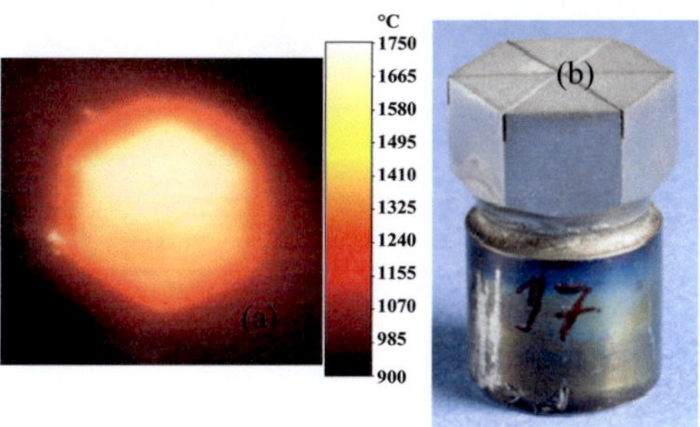

Figure 3.9-16: Second HHF test series, mock-up #17 (optimized HEMJ, castellated). (a): IR temperature image at 10 MW/m^2, mfr ~9 g/s; $T_{in, He}$ ~550 °C (T_{max} ~1700 °C).

Results:

(i) Mockup survived 89 cycles at 10 MW/m^2 w/o damage;

(ii) The experiment was terminated after detecting a temperature increase of the tile surface, no damage (b).

Good points: He loop and mockup remained absolutely intact.

Figure 3.9-17: Second HHF test series, mock-up #18 (optimized HEMJ, castellated).

Results:

Mockup survived 102 cycles at ~9.5 MW/m^2, mfr 12.5 g/s; $T_{in, He}$ ~550 °C (T_{max} ~1600 °C);

Good points: Excellent performance, no any damages, no leaks, stable surface temperature from cycle to cycle, He loop and mockup remained absolutely intact.

Mock-up no.	Cycle number at heat flux (MW/m^2)/(beam on/off)*	mfr (g/s)	T_{He} in/out (°C)	Δp (MPa) at mfr
12 (d)	18 at 10; ($\rightarrow$ gas leak at the central upper area of the tile, no significant visible damages)	9–10	560/610	0.2
13 (c)	70 at 10; ($\rightarrow$ gas leak at the central upper area of the tile, slight cracking of the tile top surface)	9	570/620	0.16
14 (c)	90 at 9; ($\rightarrow$ surface temperature increasing during cycling, tile detaching, no gas leak, melting and cracking of the tile)	9	560/610	0.17
15 (c)	Gas leak appeared between tile and conic sleeve during screening tests, no visible damages	9	550/590	0.17
20 (d)	Gas leak appeared between tile and conic sleeve during screening tests, no visible damages	9	550/590	0.17
17 (d)	89 at 10; ($\rightarrow$ experiment was terminated after detecting tile temperature increase, no gas leakage, no damages)	9	570/620	0.18
19 (d)	Gas leak between tile and conic sleeve during first heating at 450 °C and 8 MPa, cracks inside the thimble (vertical visible) and in thimble/conic sleeve brazing zone			
18 (d)	102 at 9.5/**; ($\rightarrow$ excellent performance, no damages, no leaks, stable surface temperature from cycle to cycle)	12.5	550/590	0.33

Option	W tile	W-WL10 joint	WL10-Eurofer joint
(a)	non-castellated	STEMET® 1311 brazing	Cu casting in conical lock
(b)	non-castellated	STEMET® 1311 brazing	Co brazing in conical lock
(c)	castellated	STEMET® 1311 brazing	Cu casting in conical lock
(d)	castellated	STEMET® 1311 brazing	Co brazing in conical lock

Beam on–off cycles:
* Default 30/30 s, otherwise ** soft ramp: 20 s – up, 20 s – hold, 20 s – down, 20 s – pause.

Table 3.9-2: Second HHF 1-finger test series 2007 performed (mockup details see Figure 3.9-13): Test campaign, test conditions used, and short results.

Summary of the second series of experiments:

1. The measures taken to improve mock-up brought a noticeable improvement in performance and resistance against thermal cyclic loadings. Despite only partial improvement in mock-up quality a significant increase in divertor performance was achieved. The last successfully tested mock-ups survived outstandingly more than 100 cycles under 10 MW/m^2 without any damages.

2. Nevertheless, tile temperature increase and gas leak during the load cycles were still observed in many mock-ups, but no damages were detected after experiment termination.

3. The main reasons for the high failure rate of mock-ups were identified which generally lie in:

   a) base material quality,

   b) manufacturing quality (W machining, jet holes drilling, EDM of W surfaces, etc.),

   c) overheating of the tile/thimble brazed joint leading to detachment, and

   d) induced high thermal stresses.

4. The decisive reason for the overheating of the brazed joint between tile and thimble lies in the use of brazing materials with too low melting temperature. The detachment of the components is the consequence, which leads to the melting of the tungsten tile.

Appropriate countermeasures to be taken:

o Investigation of other suitable brazing materials with higher melting point (see chapter 4.1.2) to prevent overheating of the joint.

o Developing of non-destructive tesing (NDT) methods for failure detection in raw material and manufactured components.

3.9.4   Durability Test of 1-Finger Mock-ups under Cyclic Thermal Loading at 10 MW/m^2

After the measures undertaken have proved to be effective to improve performance of the cooling finger, the maximum attainable number of cycles under a heat load of 10 MW/m^2 (hot startup and shutdown simulation) is one of the objectives of the following test series.

The third HHF test series 2008 [3.9-4] contained ten mock-ups with castellated W tiles (Figure 3.9-18). They differ in tile design, tile material used, brazing filler metal, and type of machining (EDM vs. milling/turning). The mock-up #18 was tested a second time, after having survived the last test unscathed. The mock-up #22 was tested twice repeated in the same test series after it succeeded in the first round. A new brazing material CuNi44 (Tbr = 1300 °C) was used for the W-WL10 joint in the mock-ups #24, 26 and 32, whereas the mock-up #26 was unusable after brazing. For the rest of mock-ups STEMET$^®$ 1311 (Tbr = 1050 °C) was used as before. The WL10 thimble/Eurofer steel joint is exclusively brazed with 71KHCP$^®$ (Co-based, Tbr = 1050 °C) filler metal.

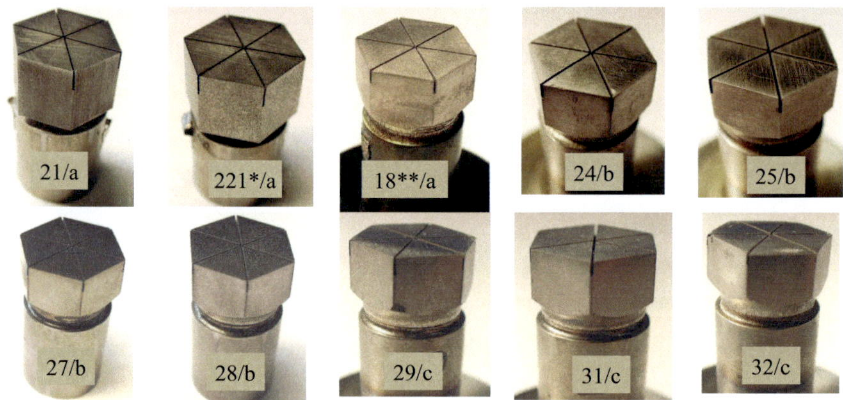

*Image after 100 cycles at 10 MW/m^2 from the previous test in the same series.
**Image after 100 cycles at 10 MW/m^2 from the previous test series (2007).

Figure 3.9-18: Third HHF test series 2008: Mock-up details.

Tungsten tile material: (a) Plansee W rod, Ø25/vertical; (b) Russian W rod, Ø30/vertical; (c) Russian rolled plate, 24 mm thick/horizontal. Type of machining tile/thimble: EDM (MU # 21&22), NC machining (MU #32), and regular turning for the rest. Brazing: W-WL10 joint: CuNi44 (MU #24,26,32), STEMET$^®$1311 (others); WL10-steel: 71KHCP$^®$ in conical lock.

The test conditions applied in the 3rd test series are as follows: (a) He mass flow rate was raised within the range of up to 13 g/s in order to keep the temperature at the tile/thimble brazing layer below T_{br} of 1050 °C; (b) Lowering He inlet temperature to a range of 450–550 °C allowed to check the functionality of mock-ups at the absorbed heat flux up to 12 MW/m^2 even with tile/thimble 'low' temperature brazing at 1050 °C; (c) Heat flux variation from 8 to 12 MW/m^2; (d) Besides standard ‚sharp ramp' (30s – on, 30s – off) some tests were partially performed with ‚soft ramp' (20s – up, 20s – on, 20s – down, 20s – pause).

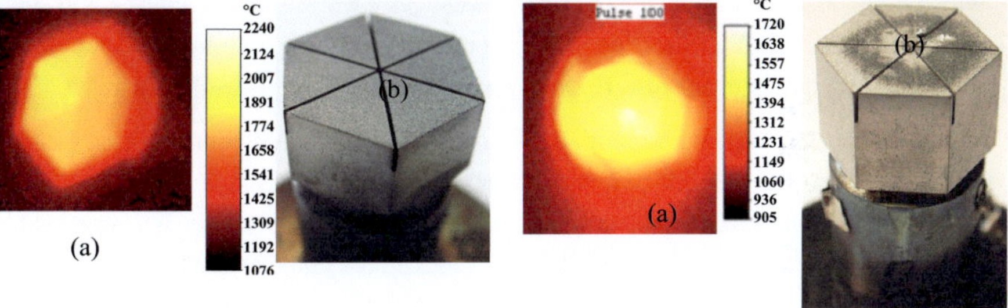

Figure 3.9-19: Third HHF test series, mock-up #18 - Test #2. (a): IR temperature image at 11 MW/m^2, mfr ~13 g/s; $T_{in,\,He}$ ~500 °C (T_{max} ~1750 °C).
Results: Mockup survived in total 112 cycles at ~11 MW/m^2, mfr 13 g/s; Tin, He ~500 °C (Tmax ~1700 °C);
Good points: excellent performance, no any visible damages, no leaks, stable surface temperature from cycle to cycle. He loop and mockup remained absolutely intact.

Figure 3.9-20: Third HHF test series, mock-up #21 (EDM). (a): IR temperature image at 10 MW/m^2, mfr ~13 g/s; $T_{in,\,He}$ ~550 °C (T_{max} ~1650 °C).
Results: (i) Mockup survived 100 cycles at ~9.5 MW/m^2 w/o damage; (ii) Slight erosion of the surface and micro cracks with small spots of melting. The reason – initial cracks in W-rod (tile-material).
Good points: Good performance, no serious damages, no leaks, stable surface temperature from cycle to cycle. He loop and mockup remained absolutely intact.

Mock-up #18 – test #2 was tested for the second time as the continuation of tests of the second series. It resisted 50 cycles at ~11 MW/m^2 under soft ramp, 50 cycles at ~11 MW/m^2 (sharp ramp), and 12 cycles at ~12 MW/m^2 (sharp ramp) without damage. The inlet temperature was decreased down to 500 °C. The surface temperature was stable from cycle to cycle. Switching from soft ramp to sharp ramp during the cycling did not show any negative results. No visible damages and no leaks occurred (Figure 3.9-19). The measured pressure loss amounts to about 0.35 MPa at

~13 g/s mas flow rate (Table 3.9-3). This value is equivalent to ~0.11 MPa for the DEMO reference case (6.8 g/s, 10 MPa, 600 °C) and agrees well with the calculated values. This mockup has an excellent performance and is <u>available for further tests</u>.

Mock-up #21 was exclusively fabricated by EDM without turning. Its tile was manufactured from the Plansee's W-rod which had initial radial oriented cracks due to fabrication process. It successfully survived 100 cycles heat load at ~9.5 MW/m^2 without damage (Figure 3.9-20). One part of the surface had a higher temperature of up to 100 K difference. This is probably due to the initial cracks in the tile material, which could also be reasons for the surface changes such as erosion, cracks, and spots (Figure 3.9-20, b). The mockup has a good performance. Its surface temperature remained constant over the test. He loop and mockup remained absolutely intact.

Mock-up #22 – test #1, also manufactured by EDM, survived in the first test run 100 cycles at 10 MW/m^2 without damage (Figure 3.9-21). It showed a good performance with stable surface temperature from cycle to cycle. No any visible damages and no leaks occured. He loop and mockup remained absolutely intact.

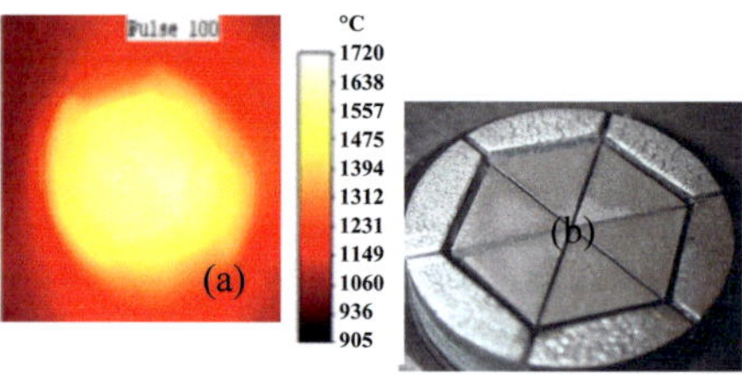

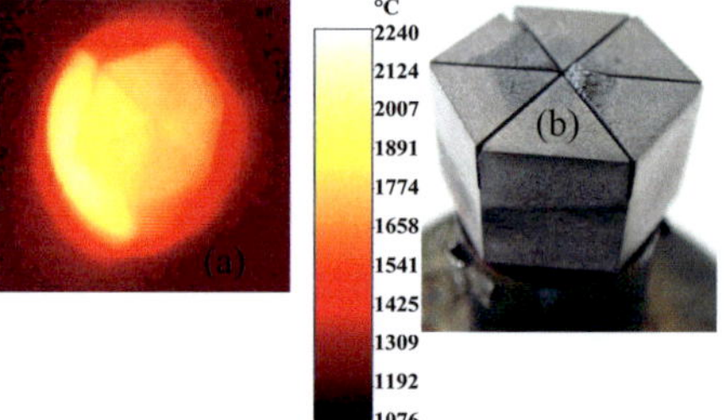

Figure 3.9-21: Third HHF test series, mock-up #22 (EDM) - Test #1. (a): IR temperature image at 10 MW/m^2, mfr ~12.5 g/s; $T_{in, He}$ ~550 °C (T_{max} ~1700 °C).
Results: Mockup survived 100 cycles at ~10 MW/m^2, mfr 12.5 g/s; $T_{in, He}$ ~550 °C (T_{max} ~1700 °C);
Good points: Good performance, stable surface temperature from cycle to cycle, no any visible damages (b), no leaks, He loop and mockup remained absolutely intact.

Figure 3.9-22: Third HHF test series, mock-up #22 (EDM) - Test #2. (a): IR temperature image at 10 MW/m^2, mfr ~13 g/s; $T_{in, He}$ ~550 °C (T_{max} ~1650 °C).
Results: (i) Mockup survived 114 cycles at ~11 MW/m^2) w/o damage; (ii) Dark spot appeared at the surface, star-shape cracks at the surface (as initial defects) are now visible.
Good points: Good performance, no any damages, no leaks, stable surface temperature from cycle to cycle. The mock-up is available for further tests.

Mock-up #22 – test #2: The inlet helium temperature was decreased down to 410 °C to have the possibility to increase the incident heat flux up to 11 MW/m^2. During the temperature cycling the inlet temperature was increased up to 550 °C. At the end of the tests the soft ramp was switched over to the sharp one, no effects were detected. This mockup survived a total of 114 cycles at ~11 MW/m^2 without significant defects (Figure 3.9-22). Only some dark spots appear on the tile surface. Star-shaped cracks that were already there from the start as initial defects had become more visible. The mock-up shows good overall performance and is <u>suitable for further testing</u>.

Mock-up # 24 has a joint between the tile and thimble brazed with CuNi44 (T_{br} = 1300°C). The basic idea was to achieve a stable operation at a heat flux of ~10–11 MW/m^2 at a minimum mfr of ~9 g/s or less. But during the load-cycling, an increase of surface temperature and a decrease in the helium temperature rise were observed (Figure 3.9-23). Such behavior with a slow cooling of the surface indicates a tile detachment or poor brazing. The mock-up survived 45 cycles at ~10 MW/m^2. Good points are that no leakage and no significant visible damages occurred.

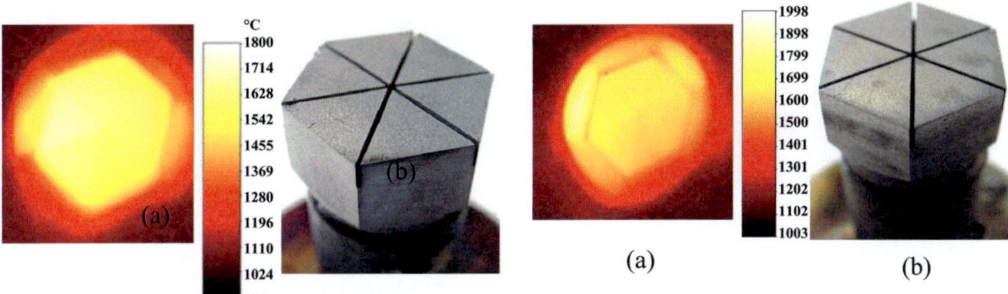

Figure 3.9-23: Third HHF test series, mock-up #24. (a): IR temperature image at 10 MW/m^2, mfr ~13 g/s; $T_{in,\,He}$ ~530 °C (T_{max} ~1600 °C).
Results: (i) Mockup survived 45 cycles at ~10 MW/m^2; (ii) Increasing $T_{surface}$ and decreasing ΔT in gas - from cycle to cycle, slow cool-down – tile detaching.
Good points: No gas leaks, no significant visible damages.

Figure 3.9-24: Third HHF test series, mock-up #25. (a): IR temperature image at 11 MW/m^2, mfr ~13 g/s; $T_{in,He}$ ~460 °C.
Results: Mockup survived 120 cycles at ~11 MW/m^2 w/o damage.
Good points: Good performance, stable surface temperature from cycle to cycle. No visible damages (b), no leaks.

Mock-up #25 is a regular mock-up type, manufactured by conventional machining methods. The inlet helium temperature was lowered to 460 °C to have the possibility to increase applied heat flux up to 11.5 MW/m^2. The tests were performed under sharp ramp. The mock-up successfully survived 10 cycles at ~10 MW/m^2, 100 cycles at ~11 MW/m^2, and 10 cycles at ~11.5 MW/m^2 without any visible damages or leaks (Figure 3.9-24). It showed good performance with stable surface temperature from cycle to cycle and is <u>available for further tests</u>.

Mock-up #27 is a regular mock-up type, manufactured by conventional machining methods. The inlet helium temperature was temporarily lowered to 470 °C to allow the tests at elevated heat flux of up to ~11.5 MW/m^2. The tests were performed under sharp ramp. The mock-up successfully survived 100 cycles at ~11 MW/m^2 and 15 cycles at ~11.5 MW/m^2 (Figure 3.9-25) without any damage and is <u>available for further tests</u>.

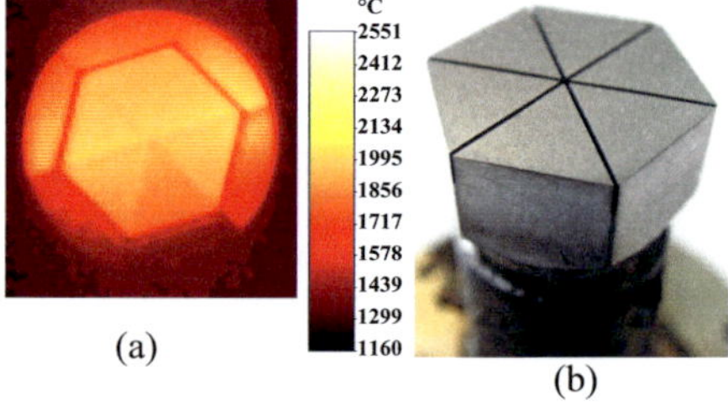

Figure 3.9-25: Third HHF test series, mock-up #27.	**Figure 3.9-26:** Third HHF test series, mock-up #28. (a): IR temperature image at 11 MW/m^2, mfr ~13 g/s; T$_{in,He}$ ~470 °C.
Results: Mockup survived 115 cycles at ~11 MW/m^2. **Good points:** Good performance, stable surface temperature from cycle to cycle. No visible damages, no leaks.	**Results:** Mockup survived 100 cycles at ~11 MW/m^2 and 12 cycles at ~12 MW/m^2 w/o damage. **Good points:** Good performance, stable surface temperature from cycle to cycle. No visible damages (b), no leaks.

Mock-up #28 is a regular mock-up type, manufactured by conventional machining methods. The inlet helium temperature was temporarily decreased down to 470 °C to have the possibility to increase applied heat flux up to ~12 MW/m^2. The tests were performed under sharp ramp. The mock-up successfully survived 100 cycles heat load at ~11 MW/m^2 and 12 cycles at ~12 MW/m^2 and had a stable temperature behavior from cycle to cycle. No leakage and no any visible damages were detected (Figure 3.9-26). With its good performance, the mockup was <u>chosen for further tests</u>.

Mock-up #29 with a tile of tungsten plate material withstood 20 cycles heat load at ~11 MW/m^2, 12 cycles at ~12 MW/m^2, and 7 cycles at ~14 MW/m^2 without damage. The inlet helium temperature was 495 °C. Due to an error in the facility, the heat flux was unstable so that the peak value increased up to ~15 MW/m^2. This overload led to a gas leak between conic ring and thimble at the end. No any visible damages were observed (Figure 3.9-27).

Mock-up #31 has a tile that is made of a tungsten plate and has a grain orientation perpendicular to the heat flux. The height of the brick is 11.3 mm instead of 12 mm, as two bricks were made from the semi-finished products with a thickness of 24 mm. The inlet temperature was lowered to 500 °C. The mock-up survived 30 cycles at ~10 MW/m^2 and 72 cycles at ~11 MW/m^2 with soft ramp. There were no visible damages and no leaks (Figure 3.9-28). This mock-up showed a good performance and had a stable surface temperature from cycle to cycle and is <u>available for further tests</u>.

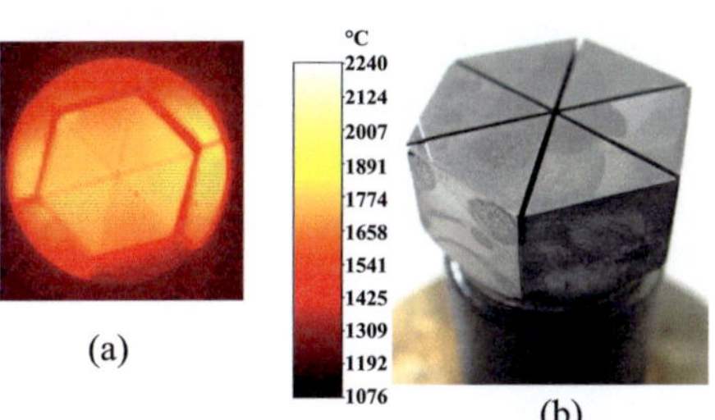

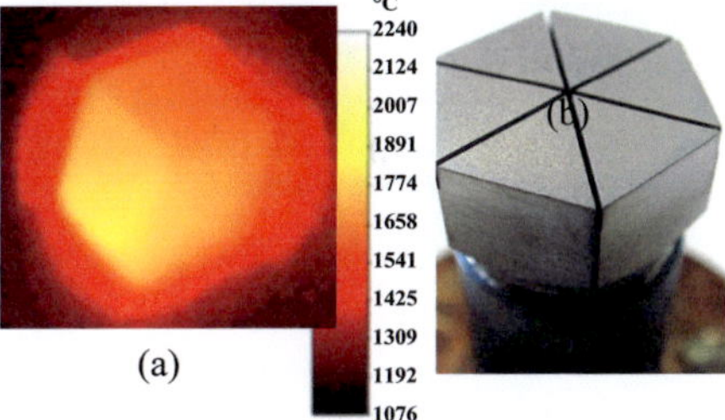

Figure 3.9-27: Third HHF test series, mock-up #29. (a): IR temperature image at 10 MW/m^2, mfr ~13 g/s; T$_{in,He}$ ~495 °C. **Results:** (i) Mockup survived 39 cycles at ~11–14 MW/m^2; (ii) Due to errors in the facility, the heat flux applied was unstable with peaks increasing up to ~15 MW/m^2, leaks occurred. **Good points:** Good performance and no leaks at 11 and 12 MW/m^2, no visible damages.

Figure 3.9-28: Third HHF test series, mock-up #31. (a): IR temperature image at 11 MW/m^2, mfr ~13 g/s; T$_{in,He}$ ~500 °C (T$_{max}$ ~1750 °C).

Results: Mockup survived 30 cycles at ~10 MW/m^2 and 72 cycles at ~11 MW/m^2 w/o damage. **Good points:** Good performance, stable surface temperature from cycle to cycle. No visible damages, no leaks.

Mock-up #32 is a test module, whose tile-thimble joint is brazed with CuNi44 alloy (T_{br}=1300°C). During the screening and cycling, a surface temperature increase and decrease in helium-temperature rise were detected. Such behavior together with slow surface cool-down is an indication for tile detaching or poor brazing. The mockup survived in total 10 cycles at ~10 MW/m^2 with a helium inlet temperature of ~550 °C. Cracks, melting and deformations of the tile surface were found (Figure 3.9-29). This suggests that the brazing with the new filler material was not yet perfect and needs further development.

Figure 3.9-29: Third HHF test series, mock-up #32. Image after tests (10 cycles at q=10 MW/m^2).
Results:
- Mockup survived < 10 cycles at ~10 MW/m^2.
- Increasing T_{surf} and decreasing ΔT in gas from cycle to cycle, slow cool-down.
- Tile detached and overheated.
- No gas leaks.

Summary of the third series of experiments:

The EDM -made mock-ups (#21, #22) show generally good performance, but no significant difference was found with regularly turned/machined mock-ups at performed testing conditions (q up to ~11 MW/m^2, cycle number up to ~200, Table 3.9-3). Mock-ups fabricated by improved machining (mechanical grinding and electrochemical grinding) show very stable performance at cyclic absorbed heat flux up to 11 MW/m^2 during more than 100 cycles. First tests with horizontal orientation of tile material structure (W plate) did not show any differences in terms of function stability of the mock-up in comparison with vertical structure (W rod) at applied testing conditions. No difference in results was detected between soft and sharp loading ramps. First tests on the mock-ups containing CuNi44 brazed joint between tile and thimble still brought no satisfactory results (cracking of the brazing interface, delamination of the tile from the thimble). This type of brazing needs further development. Finally, six mock-ups (#18, 22, 25, 27, 28, and 31) with the best performance result were chosen for further testing (Figure 3.9-30). The measured pressure loss amounts to about 0.35 MPa at ~13 g/s mas flow rate (Table 3.9-3). In general, the measured pressure losses agree well with the calculation.

Mock-up no.	Cycle number at heat flux (MW/m^2)/(beam on/off)	mfr (g/s)	T_{He} in/out (°C)	Δp (MPa) at mfr
18 (a) test #2	50 at ~11/(i); 50 at ~11/(ii); 12 at ~12/(ii); ($\rightarrow$Good performance, no any damage)	13	500/540	0.35
21(a)	100 at 9.5/(i) ($\rightarrow$Good performance, no serious damages)	13	550/590	0.35
22 (a) test #1	100 at 10/(i) ($\rightarrow$Good performance, no any damage)	13	550/590	0.35
22 (a) test #2	54 at 10.5/(i); 50 at 11/(i); 10 at 11/(ii); ($\rightarrow$Good performance, no serious damages)	13	550/590; 410/550; 410/550;	0.35
24 (b)	45 at 10/(ii) ($\rightarrow$ Tile detached, no gas leaks)	13–9	530/595	0.35@13
25 (b)	45 at 10/(ii); 100 at 11/(ii); 10 at 11.5/(ii) ($\rightarrow$Good performance, no any damage)	13	460/500	0.30
27 (b)	100 at 11/(ii); 15 at 11.5/(ii) ($\rightarrow$Good performance, no any damage)	13	470/515	0.30
28 (b)	100 at 11/(ii); 12 at 12/(ii)	13	470/520	0.30
29 (c)	20 at 11/(ii); 12 at 12/(ii); 7 at 12–14/(ii);	13	495-550	0.30
31 (c)	30 at 10/(ii); 72 at 11/(ii); ($\rightarrow$Good performance, no any damage)	13	500/540	0.35
32 (c)	10 at 10/(ii); ($\rightarrow$ Tile detached, overheating, no gas leaks)	13–11	550/590	0.35@13

(a): Plansee W rod, Ø25/vertical, (b): Russian W rod, Ø30/vertical, (c): Russian rolled plate, 24 mm thick/horizontal. Type of beam on–off cycling: (i) soft ramp 20 s – up, 20 s – hold, 20 s – down, 20 s – pause, (ii) sharp ramp 30/30 s.

Table 3.9-3: Third HHF 1-finger test series 2008 performed (mockup details see Figure 3.9-18): Test campaign, test conditions used, and short results.

KIT design,
Plansee W
rod

Tested twice,
total no. of cycles > 200:
50 @ ~11 MW/m^2 (soft
ramp)
50 @ ~11 MW/m^2 (sharp)
12 @ ~12 MW/m^2 (sharp)

RF design,
Plansee W
rod

Tested twice,
total no. of cycles > 200:
54 @ ~10.5 MW/m^2 (soft
ramp) 50 @ ~11 MW/m^2
(soft)
10 @ ~11 MW/m^2 (sharp)

KIT design,
Russian W
rod

Tested once,
Total no. of cycles > 100:
10 @ ~10 MW/m^2 (sharp
ramp)
100 @ ~11 MW/m^2 (sharp)
10 @ ~11.5 MW/m^2 (sharp)

KIT design,
Russian W
rod

Tested once,
total no. of cycles > 100:
100 @ ~11 MW/m^2
15 @ ~11.5 MW/m^2
(all sharp ramp)

KIT design,
Russian W
rod

Tested once,
total no. of cycles > 100:
100 @ ~11 MW/m^2
12 @ ~12 MW/m^2
(all sharp ramp)

KIT design,
Russian
rolled plate

Tested once,
total no. of cycles > 100:
30 @ ~10 MW/m^2
72 @ ~11 MW/m^2
(all soft ramp)

Figure 3.9-30: Six surviving mock-ups from the 3rd series chosen for further testing.

The fourth HHF 1-finger test series 2010

After a new EB gun (200 kW, 40 kV) has been installed in the test facility, the 4th HHF experiment series was started early 2010 with the tests of the six mock-ups survived from the last test series (Figure 3.9-30). The same test conditions as in the previous test series with the old gun were used. Tests were performed at a high heat flux of at least 10 MW/m^2. Tables 3.9-4 and 3.9-5 show the mock-up details and the HHF test results. The tested mock-ups survived differently between 180 and 1100 cycles under the maximum heat load of at least 10 MW/m^2 before failure. Figures 3.9-31 to 3.9-36 show the corresponding test results of all tested mock-ups after the failure. Two types of failures were identified: a) damage on top, helium leak e.g. Mock-up #18 (Figure 3.9-31) and b) damage on the side of tile, overheating, but no leak e.g. mock-up #22 (Figure 3.9-32). The best results were obtained with the optimized mock-up #18, which had survived more than 1000 cycles under 10 MW/m^2

before it failed after a total number of cycles of 1112 (Table 3.9-5). The first breakthrough was thus achieved. Presumably due to an inconsistent calibration of infrared temperature measurement now using two-color optical pyrometer method and absorbed power with the new gun running with digital beam rastering, no reliable temperature information could be supplied from these tests. In future tests 1-finger and 9-finger module mockups with improved manufacturing and related joining technologies as described above will be tested.

W tile geometry			W tile material / grain orientation	Type of fabrication
Design type	tile height (mm)	castellation depth (mm)		
KIT	12	2.7	Plansee rod/vertical	turning/grinding
RF	12	4	Plansee rod/vertical	EDM
KIT	12	2.7	RF rod/vertical	turning/grinding
KIT	12	2.7	RF rod/vertical	turning/grinding
KIT	12	2.7	RF rod/vertical	turning/grinding
KIT	11.3	2.3	RF rolled plate/horizontal	turning/grinding

Table 3.9-4: The fourth experiment series 2010 with new EB gun: Mockup details.

Mockup parts: castellated W tile, Plansee WL10 thimble, Eurofer structure; Brazing: tile/thimble with STEMET®1311, thimble/steel conic sleeve with 71KHCP, both at 1050 °C brazing temperature. Absorbed power $\geq$10 MW/m^2, Beam on/off, 15/15 s; Helium coolant: mass flow rate 13 g/s, helium inlet temperature 500 °C.

Mock-up no.	Number of cycles		Total number of cycles to failure (summed over all test series)	Failure type
	reached in the previous testseries	reached in the last tests 2010		
#18	214 (2nd and 3rd)	900	1114	A
#22	214 (2nd and 3rd)	50	264	B
#25	120 (3rd)	300	420	B
#27	115 (3rd)	299	414	B
#28	112 (3rd)	99	211	B
#31	102 (3rd)	74	176	A

A: damage on top, helium leak; B: damage on the side of tile, overheating, no leak.

Table 3.9-5: The fourth experiment series 2010 with new EB gun: HHF test results.

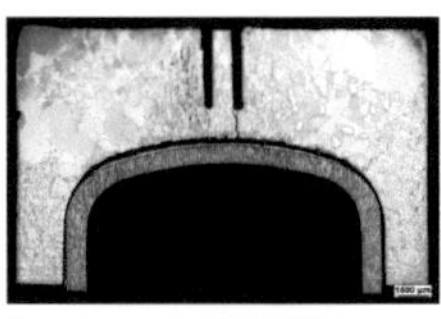

[courtesy of FZJ]

(Image before testing)

Figure 3.9-31: Fourth HHF test series, mock-up #18 - Test #3.

Results: (i) Mockup survived 900 cycles at ≥ 10 MW/m^2, total 1114 cycles achieved; (ii) Dark spot appeared at the surface, helium leak damage on top.

Figure 3.9-32: Fourth HHF test series, mock-up #22 - Test #3.

Results: (i) Mockup survived 50 cycles at ≥ 10 MW/m^2, total 264 cycles achieved; (ii) Surface temperature increased during the experiment. Cracks in the tile without gas leakage.

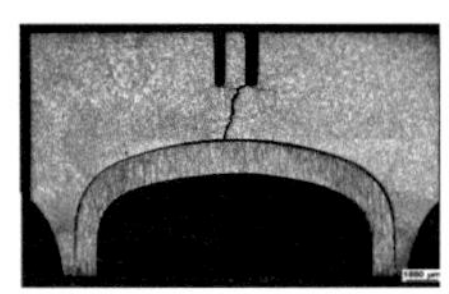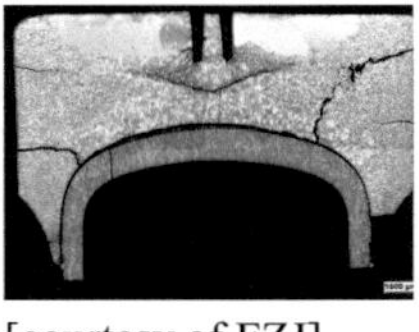

[courtesy of FZJ]

[courtesy of FZJ]

Figure 3.9-33: Fourth HHF test series, mock-up #25 - Test #2.

Results: (i) Mockup survived 300 cycles at ≥ 10 MW/m^2, total 420 cycles achieved; (ii) Surface temperature increased during the experiment. Cracks on the tile flanks and in the center of the tile at the bottom of the groove without gas leakage.

Figure 3.9-34: Fourth HHF test series, mock-up #27 - Test #2.

Results: (i) Mockup survived 299 cycles at ≥ 10 MW/m^2, total 414 cycles achieved; (ii) Cracks on the tile flanks without gas leakage. Melted area on tile surface.

 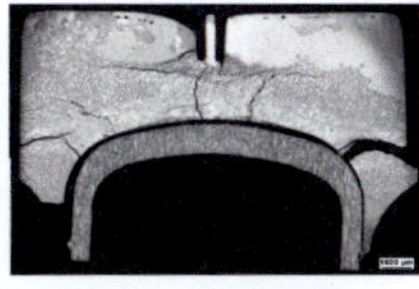

[courtesy of FZJ]

Figure 3.9-35: Fourth HHF test series, mock-up #28 - Test #2.
Results: (i) Mockup survived 99 cycles at ≥10 MW/m^2, total 211 cycles achieved; (ii) Melting of mock-up surface. No leak.

Figure 3.9-36: Fourth HHF test series, mock-up #31 - Test #2.
Results: (i) Mockup survived 74 cycles at ≥10 MW/m^2, total 176 cycles achieved; (ii) Cracks on the tile flanks without gas leakage. Gas leakage in the center of the top surface of the tile.

3.9.5 Summary of the High Heat Flux Test Results

The aim of the HHF experiments is the design verification and proof of principle. Despite the limited number of available mockups reasonable experimental results have been obtained by a balanced parameter variation. The functionality of the design and cooling ability with helium was quickly confirmed in the beginning of the experiment. Figures 3.9-37 and 3.9-38 show the corresponding bar graphs of the achieved maximum performance and the total number of cycles of the tested mock-ups. About 97 % of the total cycles were reached at a high heat flux of 9–14 MW/m^2. It can be seen from this well that the required performance of 10 MW/m^2 was met from the beginning. However, the number of cycles reached at this power level at the beginning was still relatively low. This is due to the poor quality of the raw material and the manufacture of tungsten parts. It was quickly realized that the cracks in the starting material and the micro-cracks induced by inadequate manufacturing tungsten parts affect the lifetime of the cooling finger strongly. It was also identified that high thermal stresses cause stress cracks. After a gradual improvement in the quality of mockup manufacturing, the rejection rate of the mock-ups due to production has become smaller, as the decreasing number of non-testable mockups in the diagram is clearly seen. Together with an optimization of the mock-up geometry, the achieved number of cycles of the tested mock-ups increased during the experiments significantly, reaching in the last series of experiments over a width of some hundred cycles, and even a peak of larger than one thousand cycles (Mock-up No. 18, optimized reference design). This trend shows that the focus of development was in the right direction, namely in the manufacturing and related joining technology of tungsten components. It became clear that the use of tungsten materials is not easy. Therefore, more stringent standards for quality control of basis materials as well as non-destructive testing of the assembled finger are essential. The obligatory rule for the W component manufacturing is the achievement of micro-crack free component surfaces. Resistance to preferred directions of crack propagation promises a homogeneous and uniform distribution of grains in the tungsten material. This could be achieved e.g. through an optimal combination of a powder injection molded tungsten tile with a deep-drawn WL10 thimble (see chapter 4.1-3). However, such fingers must first be subjected to the HHF tests.

In summary, the test results confirm good functionality of the divertor design, including the thermal-hydraulic and thermo-mechanical behaviors of the finger module system. The required durability of the divertor at 100–1000 thermal cycles (chapter 3.2) has been achieved. The failure of the mock-ups occurred is not due to the fatigue of the components but to the joining and fabrication methods.

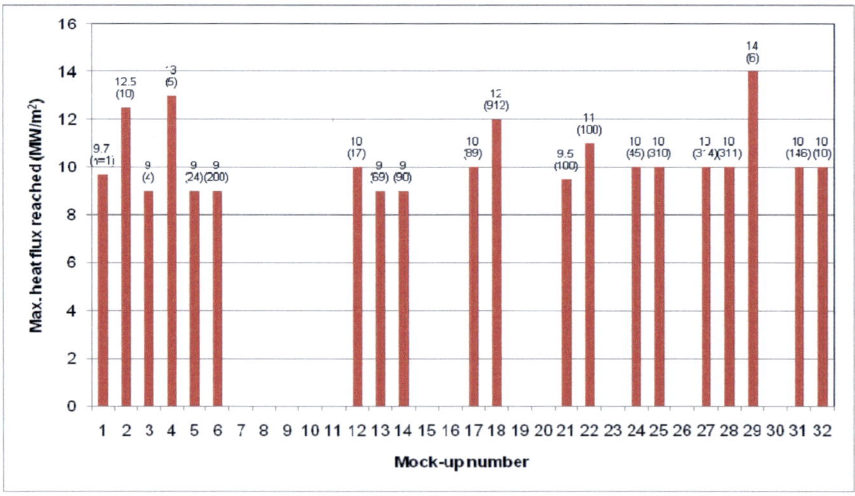

Figure 3.9-37: Absolute maximum heat flux (MW/m^2) achieved by the tested mockups. The values in parentheses are the corresponding number of cycles (n) at this load.

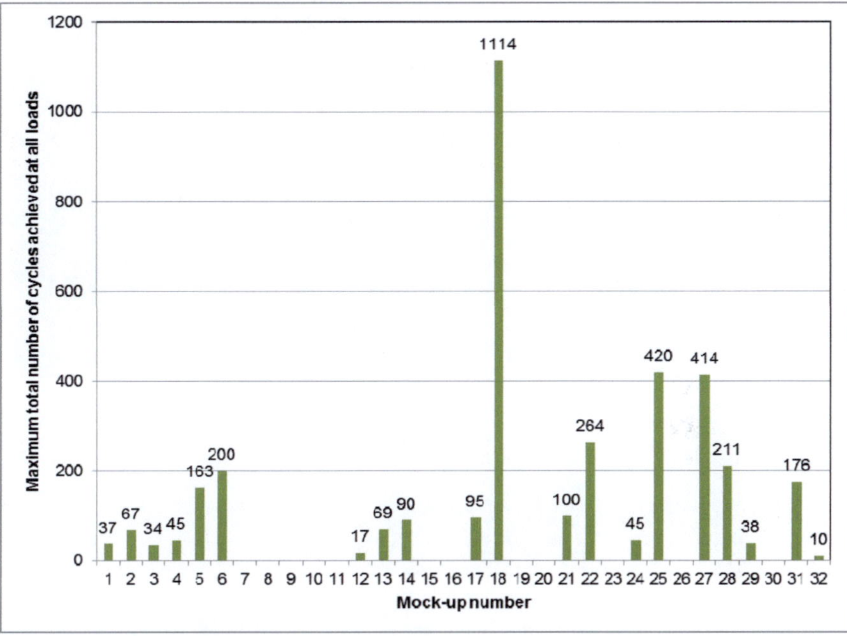

Figure 3.9-38: Total number of cycles achieved by the tested mock-ups, about 97 % of it at a heat load of 9–14 MW/m^2.

3.9.6 Manufacturing and Thermohydraulic Tests of 9-Finger Steel Mock-Up

Based on the KIT 9-finger module design (Figure 3.9-39, a) a small manufacturing-oriented modification was made by EFEREMOV. Thereafter, the individual components of steel were produced there in the conventional manner and assembled for a complete test module with interface ports (Figure 3.9-39, b–e). The main assembly steps are: HIPing steel body parts, fixing the cartridge in steel body, brazing of the finger parts (tile, thimble, transition piece and steel body), and electron beam seal welding of the last connection. The 9-finger steel mock-up was then mounted to the helium loop (Figure 3.9-39, f).

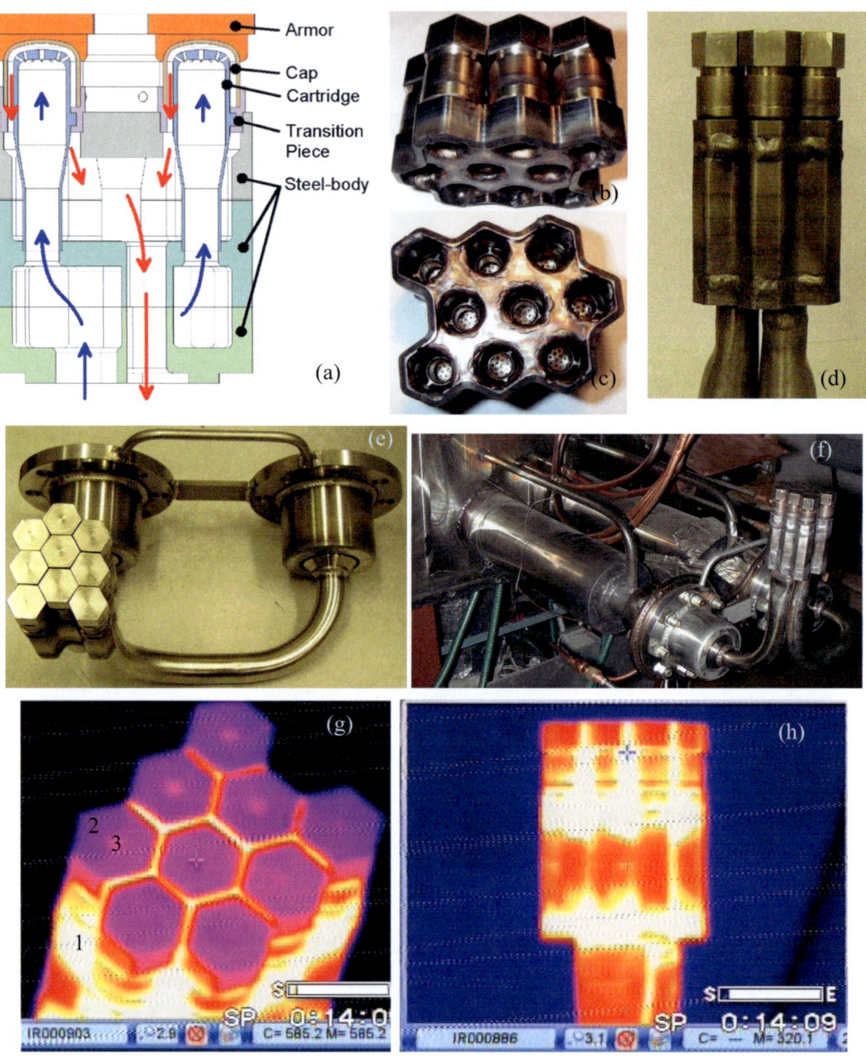

Figure 3.9-39: 9-finger steel mock-up for thermohydraulic tests.

(a) Basis HEMJ 9-finger module design,
(b, c) assembling (TIG welding) of the upper part of the module (top and bottom view),
(d) final assembly and TIG welding of the module parts,
(e) complete 9-finger mock-up unit after final assembly,
(f) 9-finger module installation to the helium loop,
(g, h) infrared images of the 9-f module during tests at T_{He} ~600 °C, $T_{surface,\ max./min}$ ~550/500 °C.

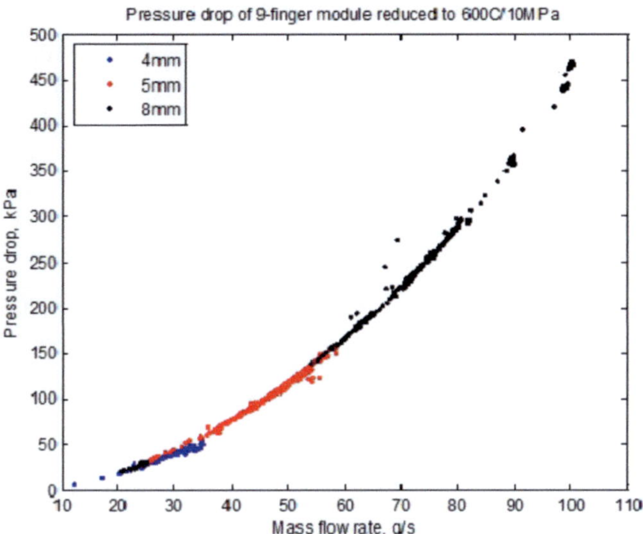

Figure 3.9-40: Normalized pressure drop measurement of 9-finger module to the DEMO design conditions (600 °C, 10 MPa).

Goal of the first gas-flow tests with a 9-finger module in 2008 without surface heat flux is the study of: a) the uniformity of flow distribution in the cooling fingers by means of the temperature distribution measurement, b) the mechanical stability of the module under internal pressure and temperature of 600 °C, and c) gas flow parameters. First thermo-hydraulic tests were performed under the conditions: a) He 600 °C, 10 MPa, b) mass flow rate variable within a range of 20 – 100 g/s by gas puffing, using 3 sizes of throttle, c) measurement of helium parameters and mass flow pulses within a time period of about 50 s, and d) measurement of surface temperature by means of an infrared camera. Three series of gas-puffing experiments were performed with three different Flow Rate Throttles (4, 5 and 8 mm). Examination of the flow distribution in the fingers via surface temperature distribution shows very uniform distribution (Figure 3.9-39, g and h). The tile surface temperatures range from about 500 °C (point 2) to about 550 °C (point 3), while the maximum temperature at the steel case is about 600 °C (point 1). The pressure loss equivalent for the DEMO reference case (9 x 6.8 g/s = 61.2 g/s) amounts to about 0.17 MPa

(Figure 3.9-40) which lies in the range predicted by CFD calculations. Only a slight increase of the height at central finger of about 0.2 mm was observed (possibly due to the bending of the upper plate). All other dimensional changes are less than 0.1 mm. These test results have confirmed the feasibility of manufacturing and the functionality of the 9-finger module well. Future HHF tests on 9-finger modules will be performed with real tungsten fingers, which will be selected from the previous one-finger tests.

4   Technological Study on High-Quality Manufacturing of Divertor Components

For a functioning design close links between the design related R&D areas are indispensable. These are materials, manufacturing technologies with special areas of mass production, high-heat-flux (HHF) tests and subsequent post examinations.

Especially in the field of development of tungsten materials for divertor applications under the EFDA-research program to date are the difficulties and problems well known and identified [4-1]. There are two types of applications for these materials that require very different properties (see also chapter 3.6.2): one for their use as plasma facing armor or shield component, the other is for structural applications. An armor material requires high crack and sputtering resistance under extreme thermal operation conditions, while a structural material must remain ductile within the operating temperature range. Both types of material have to be stable with respect to the high neutron doses and helium production rates. The development of a structural material for the divertor is considered the most critical issue. Developing low-activation brazing materials is still a problem. A complete picture of the irradiation performance of tungsten materials is not yet available.

4.1   Machining of Tungsten and Tungsten Alloy Parts

As mentioned above in chapter 3.6.2, tungsten has been selected as divertor material due to its excellent material properties such as high thermal conductivity, high strength and high sputtering resistance. On the contrary, its high hardness (460 HV 30) and high brittleness make the fabrication of tungsten components comparatively difficult. From earlier experiments (see chapter 3.9), it was recognized that micro-cracks on the surface of tungsten parts and excessive temperature at the braze joint are the main reasons for the shortened life time of the divertor cooling finger. This is especially the case when the finger is subjected to temperature cyclic loading. The micro-cracks of a depth of about 30–50 μm [4.1-1] (Figure 4.1-1) were found to be initiated by EDM (electro discharge machining) and/or conventional machining (turning, milling, grinding) with insufficient surface quality. They lead to crack growths in tile and thimble during thermal cyclic-loading. This is the motivation for this work with the aim to improve the quality of machining tungsten parts. Generally, requirements on high accuracy and excellent surfaces are important for reaching high performance, high reliability, and high functionality of the divertor.

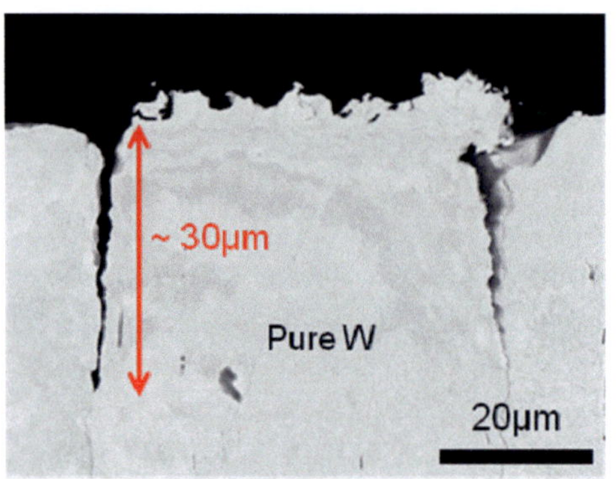

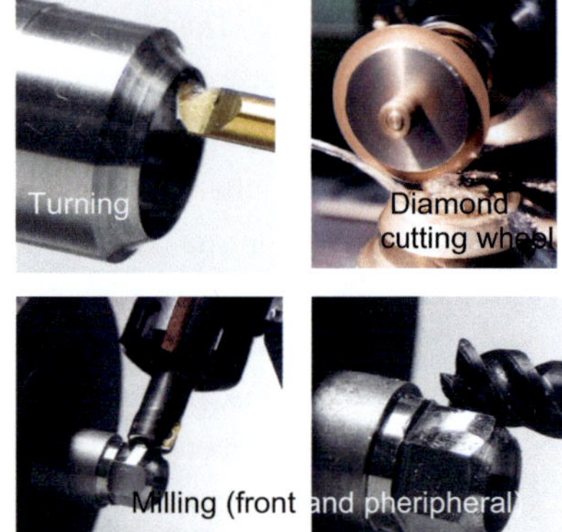

Figure 4.1-1: EDM induced micro cracks into the tungsten surface (as machined, not loaded).

Figure 4.1-2: Machining methods for bulk tungsten tile.

A detailed study of the tungsten part machining [4.1-2] was started at KIT in 2008. The investigation involves turning and milling of tungsten components (W tile and WL10 thimble) on a universal machine center (Traub TNA 300), which enables both turning and milling without any repositioning, as well as on a milling machine (DMU 50 eVolution). The latter offers more options and flexibility, e.g. higher number of revolutions, possibility for performing dry milling (i.e. milling without liquid cooling), large number of tools (36), flexible clamp as well as 5-axis machining techniques. For the assessment Plansee's deformed W-rod Ø25 mm for tile and deformed WL10 rod Ø21.5 mm for thimble machining, respectively, were used. Various parameters, such as cutting speed, feed rate, etc. were varied. In another study, different processes (turning and milling) for the production of tungsten tiles were compared. The hexagonal flanks as well as the top plasma facing surface of the tile can be machined by either front or peripheral milling (Figure 4.1-2). The overall results yield: a) generally, excellent micro-crack free surface quality was achieved by both turning and milling (Figure 4.1-3, top), b) dry milling has an in self-removal of most of the frictional heat by flying chip which helps reduce cutting tool wear when compared to turning, c) milling was found to be optimal procedure for W tile production since it offers shorter processing time by a factor of 4 than turning, d) for the hexagon contour circular front milling is recommend due to its higher accuracy, whereby machining the top surface to be carried out in a "from the edge to the center" manner to avoid break out of the edges. While machining W tile is a challenge, machining WL10 thimble is not a problem because it has a simple cylindrical shape suitable for turning. Figure 4.1-3, bottom illustrates the individual parts of a complete cooling finger (tile, thimble and conic sleeve) manufactured in such a way with high quality. They are prepared for assembly and following HHF tests at Efremov (chapter 3.9.2).

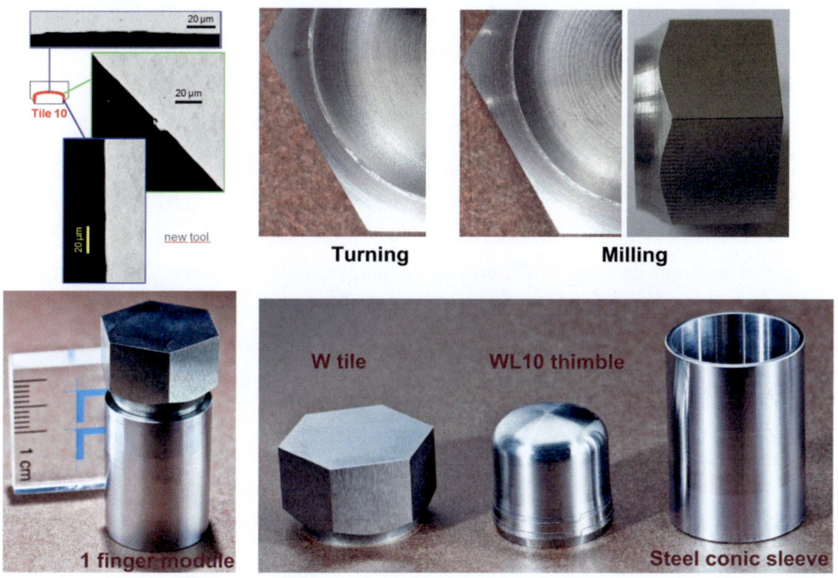

Figure 4.1-3: Crack-free surfaces of tungsten parts achieved by machining (top). W tile, WL10 thimble, and steel conical sleeve manufactured at KIT (bottom) [4.1-2]. Image courtesy of J. Reiser.

4.2 Joining of Mock-up Parts

Welding is not applicable due to the problems concerning grain growth and other microstructural changes of the W and ODS alloys during joining. High-temperature brazing and diffusion bonding are considered alternative methods. As shown in figures 5.5-8 and 6.1-4, the plasma facing surface of tungsten tile has a hexagonal cross-sectional shape with a width across flats of 18 mm. The lower surface of the tile is a concave shape that fits exactly to the thimble head shape and forms a stable braze joint. There are two different types of braze joints: a) the connection between the tungsten tile and the WL10 thimble and b) the connection between the WL10 thimble and steel structure, which is shown here as a transition piece to the base plate in form of a conical sleeve (Figure 4.2-1). In collaboration with Efremov, the first studies of tungsten brazing were carried out in 2003 [3.9-1]. STEMET®1311, an amorphous alloy (Ni based,16.0Co, 5.0Fe, 4.0Si, 4.0B, 0.4Cr, composition in wt.%), brazing temperature T_{br} = 1050 °C, was initially chosen as filler material for the upper W-WL10 braze joint and cast copper (melting point: 1083 °C) for the lower WL10-steel braze joint. Alternative for the latter: brazing with 71KHCP® (Co-based, 5.8Fe, 12.4Ni, 6.7Si, 3.8B, 0.1Mn, P≤0.015, S≤0.015, C≤0.08, composition in wt.%), T_{br} = 1050 °C. After initial difficulties, such as voids at the curved surface between the W tile and WL10 thimble error-free brazing was succeeded by the use of thin, star-uniformly distributed 40 µm brazing foil strips.

A type of failure observed in the course of the preceding tests was the detachment of tile and thimble due to an overheating of the brazed joint – top surface melting of the W tile as a consequence – when ramping up the incident heat flux beyond 13 MW/m². This failure was assumed to be caused by overheating of the W tile/WL10 thimble joint brazed with STEMET®1311. In order to improve the braze joint a study on new brazing technology for high-temperature brazing has been launched at KIT [4.2-1]. New brazing filler 60Pd40Ni (liquidus temperature T_{liq} = 1238 °C) was chosen for the W-WL10 joint (working temperature ~1200 °C), taken into account the recrystallization temperature of WL10 material (1300 °C) (Figure 4.2-1). For the brazing of WL10-Steel joint (working temperature ~700 °C) 18Pd82Cu filler (T_{liq} = 1100 °C) was found suitable. A common muffle furnace was used which allows for 10^{-5}–10^{-4} mbar vacuum and a homogeneous temperature distribution. Preparation steps are sand blasting and acetone ultrasonic bath. In both cases W-WL10 joint with PdNi and WL10-steel joint with CuPd good adhesion to the base material of the parts were achieved. Figures 4.2-2 and 4.2-3 show the EDX scan results of the two successful brazed joints [4.2-2]. In the EDX spectra (bottom) the EDX signal intensity is plotted as a function of photon energy corresponding to the point-scan data of the elements in the table (top).

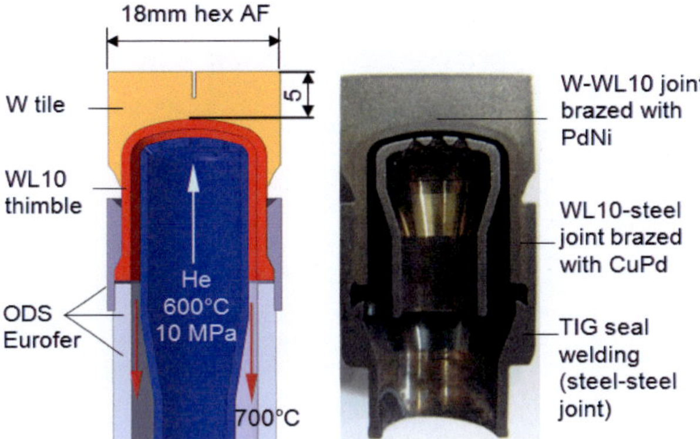

Figure 4.2-1: Reference design HEMJ, left: structure of 1-finger module, right: completely fabricated finger with brazed joints.

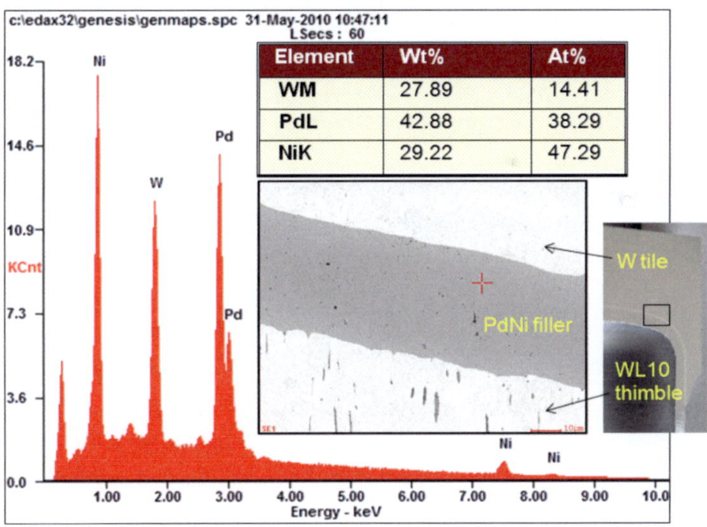

Element	Wt%	At%
WM	27.89	14.41
PdL	42.88	38.29
NiK	29.22	47.29

Figure 4.2-2: SEM and EDX scan results of a successful brazed joint W tile - WL10 thimble with PdNi40 [4.2-2]. Image courtesy of L. Spatafora and M. Müller.

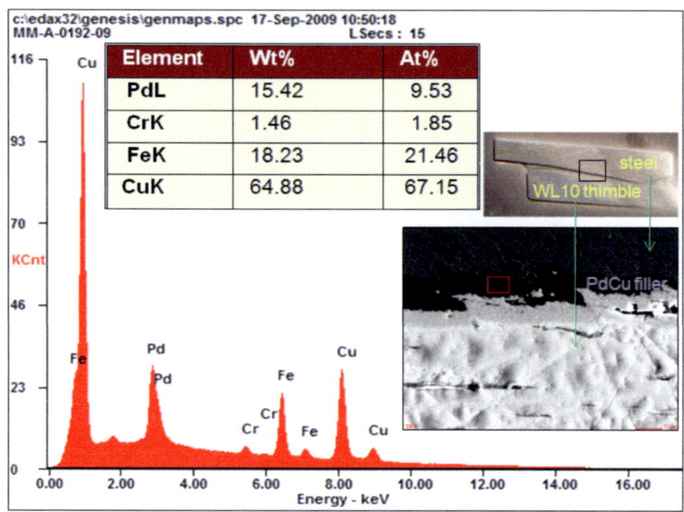

Element	Wt%	At%
PdL	15.42	9.53
CrK	1.46	1.85
FeK	18.23	21.46
CuK	64.88	67.15

Figure 4.2-3: SEM and EDX scan results of a successful brazed joint WL10 thimble - steel conic sleeve with PdCu [4.2-2]. Image courtesy of L. Spatafora and M. Müller.

4.3   Mass Production Process for Divertor Components

Because of the required large number of divertor cooling fingers of about 300,000 for the entire reactor a cost-effective method for mass production of tungsten parts is of great advantage. For the economic manufacture of functional and load-oriented divertor components from tungsten material, two methods have been investigated at KIT. These are powder injection molding (PIM) of tungsten tile and deep drawing of tungsten alloy thimble.

Tungsten Powder Injection Molding: In general, PIM is a near net shape process for the manufacturing of high volume high precision components that is widely used in industry. The advantage of this method is in addition to the cost and time savings in the fact that no preferred direction of the grain orientation is to be expected in such injection molded material. Thus, the risk of longitudinal cracking of the tile surface to the thimble head is lower than in tile which is manufactured from forged tungsten rods. This typical kind of cracks was often identified in the post-examination of HHF tested mock-ups [4.1-1] that have been conventionally produced from solid material.

A special application of PIM to the mass production of tungsten tiles is reported in [4.3-1] in detail. Key steps are feedstock formulation, injection moulding process itself, debinding, and a combined compacting process sintering plus hot isostatic pressing (HIP). The feedstock preliminary investigation showed that, with regard to the better flow properties, a binary powder system, a mixture of two particle sizes 50 wt.-% W1 (0.7 µm Fisher Sub-Sieve Size (FSSS)) and 50 wt.-% W2 (1.7 µm FSSS) proves to be optimal for this application (Figure 4.3-1). The first PIM results are very promising. For example, a compacted density of the product of almost 98.6–99 % of the theoretical density with a grain size of about 5 µm and a Vickers-hardness of 457 HV0.1 after sintering and HIP steps have been achieved.

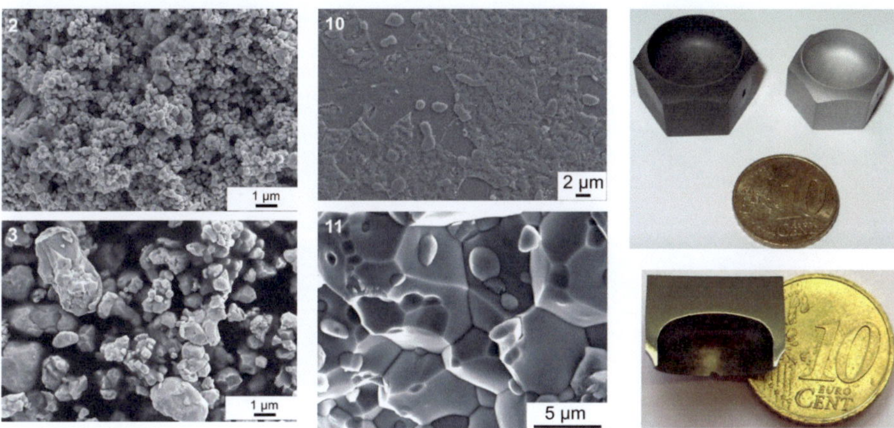

Figure 4.3-1: PIM of W tile. Left: the starting material, binary powder particle sizes 0.7 µm FSSS (2) and 1.7 µm FSSS (3); right: the finished product highly compacted tile with high density without any voids [4.3-1]. Image courtesy of S. Antusch.

Tungsten Deep Drawing: For mass production of WL10 thimble which is a structural part of the divertor a forming process deep drawing is being investigated. This kind of forming process provides an advantage in that the grains of the material are formed uniformly along the contour, which is favorable for the strength increase in the structure. First forming tests were performed with press-rolling method [4.1-2] on steel and TZM sheets which were heated to a working temperature of 400 °C by using butane gas flame (Figure 4.3-2, left). A higher temperature was not used to avoid oxidation of tungsten. During processing, the temperature was measured using a pyrometer. Figure 4.3-2, right shows good results of roll-pressed thimbles of steel and TZM materials without failure.

Figure 4.3-2: Roll pressing attempt using butane gas heating (left), thimble cap from 1 mm sheet of steel and TZM (right) [4.1-2]. Image courtesy of J. Reiser.

Figure 4.3-3: Successful deep-drawing tests with 1 mm W-sheet. Left: Vacuum furnace (courtesey V. Toth/KIT); Top right: newly developed tool made of tool steel; Bottom right: two deep-drawn W thimble-like caps (Ø15 x 1) of about 8 and 11 mm height, respectively.

In a further step cupping was performed on 1 mm W sheets in a newly constructed tool (Figure 4.3-3, top right). A vacuum furnace was used (Figure 4.3-3, left). In the first experiment, thimble-like W-caps (Ø15 x 1) of about 8 and 11 mm height were successfully deep drawn (Figure 4.3-3, bottom right). Here, a working temperature of 700 °C and a maximum force of up to about 20 kN were applied. The next step will be the manufacturing of tungsten thimble in its original geometry by means of deep-drawing process.

5   Summary and Outlook

Since the discovery of the H-mode in the ASDEX experiment called "divertor I" in 1982, the divertor, due to its excellent plasma insulation and cleaning properties, has become an integral part of all modern tokamaks and stellarators, not least the ITER machine. From a technological perspective, developing a divertor is a big challenge due to the diverse requirements to be met. One of the most important of these is to resist a very high heat load of at least 10 MW/m^2. In the course of the EU PPCS, different divertor types (WCD, HCD, and LiPb-cooled divertor) were investigated. The choice of divertor type is primarily governed by the desire to use the same coolant type as for the blanket. Additionally, operation with a high coolant exit temperature is particularly important for a power plant in order to achieve a high thermal efficiency in the power conversion system. In the first PPCS stage between 1999–2001 basic concepts of the above mentioned divertor types were studied. Their design principles, advantages and disadvantages, and analytical results are outlined in chapter 2.3.

Helium-cooled divertor designs have been favoured by most power plant models because of the chemical and neutronic inertness of helium, also allowing operation at considerably higher temperatures and lower pressures than water-cooled divertors. During the early development stages (1999–2001), the theoretical performance limit of the HCD plate designs was successively increased from 5 MW/m^2 to 10 MW/m^2 using various cooling techniques (chapter 2.3.3). One of the resulting crucial items besides high thermal loads are high thermal stresses encountered in the continuous plate design, resulting from suppressed bending of the plate structure by a strong mechanical support.

Based on the mentioned requirements and in order to meet the existing challenges a new design for a He-cooled divertor could be realized, showing the following advantages compared with former designs:

o   Reduction of local thermal stresses by a modular divertor design

o   Realization of the required cooling rates by helium jet impingement cooling

o   High thermal performance, simple construction and easy fabrication by modular cooling finger design o   High thermal resistance and good thermal conductivity through the use of tungsten-based materials were necessary

Detailed design and fabrication studies (chapters 3 and 4) as well as HHF experiments were carried out in a combined testing facility (TSEFEY EB device and moveable He loop) at Efremov for verification of the design and proof of principle (chapter 3.9). The latest experimental results already confirm the divertor's ability to accommodate a heat load of up to 14 MW/m^2, well above the design target of 10 MW/m^2. A maximum number of cycles of more than 1100 at a heat load of 10 MW/m^2 was achieved well beyond the target of 1000.

The developed divertor concept proposed for a fusion power plant to be built beyond ITER has demonstrated its principal feasibility and functionality and hence the used design process and tools can be conceived as verified and validated. It was thus an important stepping stone provided for further R&Ds towards mature power plant application, particularly in the areas of materials, fabrication and irradiation.

Nevertheless, a large effort still has to be spent to improve the design in terms of robustness against thermo-mechanical load cycling to enhance its lifetime. Intermediate-term R&D issues include: Development of mass production and non-destructive testing methods for divertor components, further development of a suitable divertor structural material with an operating window in the range 600–1300 °C, irradiation experiments of structural materials in typical neutron environments of fission and of the presently designed intense fusion neutron source IFMIF (International Fusion Materials Irradiation Facility) with DEMO-relevant neutron fluence, as well as completion of a divertor test module (TDM) to be proposed in the ITER which is to be tested first in a bigger helium loop e.g. HELOKA (Helium Loop Karlsruhe).

A List of Figures and Tables

B   References

[1-1]   IPCC   Fourth   Assessment   Report: Climate   Change   2007. http://www.ipcc.ch/publications_and_data/publications_ipcc_fourth_assessment_repo rt_wg1_report_the_physical_science_basis.htm]. Retrieved July 7, 2010.

[1-2] UN World Population Prospects, 2010 Revision, retrieved July 7, 2010, from http://esa.un.org/unpd/wpp2008/all-wpp-indicators_components.htm.

[1-3] W. Marth, The SNR 300 Fast Breeder in the Ups and Downs of its History, Forschungszentrum Karlsruhe, Wissenschaftliche Berichte, KfK 5455, 1994.

[1-4] Energie in Deutschland, Bundesministerium für Wirtschaft und Technologie (BMWI), April 2009, retrieved July 20, 2010, from
http://www.energie-verstehen.de/Dateien/Energieportal/PDF/energie-in-deutschland,property=pdf,bereich=energieportal,sprache=de,rwb=true.pdf.

[1-5] REGIERUNGonline - Energiegewinnung auf neue Füße stellen, Magazin für Wirtschaft und Finanzender Bundesregierung, Nr. 079/02-2010.
http://www.bundesregierung.de/Content/DE/Magazine/MagazinWirtschaftFinanzen/0 79/t4-energie-energiegewinnung-auf-neue-fue_C3_9Fe-stellen.html.   Retrieved   July 26, 2010.

[1-6] G. Janeschitz, Lecture Notes "Einführung in die Fusionstechnologie", Karlsruhe Institute of Technology (KIT), 2005.

[1-7] R. Klingelhöfer, Basic Physics of Nuclear Fusion, Lecture Notes, Karlsruhe Institute of Technology (KIT), 1987.

[1-8] http://de.wikipedia.org/wiki/Coulombwall. Retrieved July 29, 2010.

[1-9] W. Maurer, Physikalische Grundlagen der Kernfusion, Kernfusion, Forschung und Entwicklung, Kernforschungszentrum Karlsruhe, 1991.

[1-10] H.-J. Hartfuss, Lecture Notes "Diagnostik von Hochtemperaturplasmen", Max-Planck-Institut für Plasmaphysik, Ernst-Moritz-Arndt–Universität Greifswald, 2004.

http://www.physik.uni-greifswald.de/hartfuss/kap01.pdf. Retrieved August 5, 2010.

[1-11] ITER - the way to new energy, http://www.iter.org/, retrieved March 2, 2011.

[1-12] R. Wolf, Magnetisch eingeschlossene Fusionsplasmen auf dem Weg zu einer neuen Energiequelle, DPG-Tagung, 21. März 2006, München.

[1-13] G. Janeschitz, ITER - the essential Step towards a Fusion Reactor, DPG-Tagung, 21. März 2006, München.

[1-14] M. Kaufmann, Einführung in die Plasmaphysik, IPP: Max-Planck-Institut für Plasmaphysik.
http://www.ipp.mpg.de/~Wolfgang.Suttrop/Vorlesung/BT98_99/loesung11.pdf.
Retrieved July 26, 2010.

[1-15] http://de.wikipedia.org/wiki/Divertor, retrieved July 26, 2010.

[1-16] Max-Planck-Institut für Plasmaphysik, 50 Jahre Forschung für die Energie der Zukunft, Max-Planck-Institut für Plasmaphysik, Garching und Greifswald, 2010, ISBN 978-3-00-031750-7.
http://www.ipp.mpg.de/ippcms/de/pr/institut/geschichte/fusion_1960_2010.pdf, retrieved August 8, 2010.

[1-17] S. Günter, K. Lachner, Wie bändigt man heißes Plasma?, Physik in unserer Zeit, Volume 38, Issue 3 (p 134-140) 2007, Wiley-VCH Verlag GmbH & Co. KGaA, Weinheim, 2007.

[1-18] www.dpg-verhandlungen.de/2010/hannover/p13.pdf. Retrieved August 10, 2010.

[1-19] R. Neu, Wolfram als Wandmaterial im Tokamak ASDEX Upgrade, Max-Planck-Institut für Plasmaphysik, Jahrbuch 2004.
http://www.mpg.de/bilderBerichteDokumente/dokumentation/jahrbuch/2004/plasmap hysik/forschungsSchwerpunkt1/index.html. Retrieved July 26, 2010.

[1-20] A. Loarte et al., Expected energy fluxes onto ITER Plasma Facing Components during disruption thermal quenches from multi-machine data comparisons, Proceedings of the 20th IAEA Fusion Energy Conference, Vilamoura, Portugal, 1-6 Nov 2004, ISBN 92-0-100405-2.

[1-21] A. Hassanein, D.L. Smith, Thermal response of substrate structural materials during a plasma disruption, Journal of Nuclear Materials 191-194 (1992) 503-507.

[1-22] G. Janeschitz, L. Boccaccini, W.H. Fietz, T. Ihli, A. Moeslang, P. Norajitra, et al., Development of fusion technology for DEMO in Forschungszentrum Karlsruhe, Fusion Engineering and Design 81 (2006) 2661–2671.

[1-23] P. Norajitra, S. I. Abdel-Khalik, L. M. Giancarli, T. Ihli, G. Janeschitz, S. Malang, I. V. Mazul, P. Sardain, Divertor conceptual designs for a fusion power plant, Fusion Engineering and Design 83 (2008) 893–902.

[2.1-1] D. Maisonnier, I. Cook, P. Sardain, L.V. Boccaccini, L. Di Pace, L. Giancarli, P. Norajitra and A. Pizzuto, DEMO and fusion power plant conceptual studies in Europe, Fusion Engineering and Design 81 (2006) 1123–1130.

[2.1-2] P. Sardain, B. Michel, L. Giancarli, A. Li Puma, Y. Poitevin, J. Szczepanski, et al., Power plant conceptual study—WCLL concept, Fusion Engineering and Design 69 (2003) 769–774.

[2.1-3] S. Hermsmeyer, S.Malang, U. Fischer, S. Gordeev, Layout of the He-cooled solid breeder model B in the European power plant conceptual study, Fusion Engineering and Design 69 (2003) 281–287.

[2.1-4] P. Sardain, D. Maisonnier, L. Di Pace, L. Giancarli, A. Li Puma, P. Norajitra, et al., The European power plant conceptual study: helium-cooled lithium–lead reactor concept, Fusion Engineering and Design 81 (2006) 2673–2678.

[2.1-5] P. Norajitra, L. Bühler, A. Buenaventura, E. Diegele, U. Fischer, S. Gordeev, E. Hutter, R. Kruessmann, S. Malang, A. Orden, G. Reimann, J. Reimann, G. Vieider,

D. Ward, F. Wasastjerna, Conceptual design of the dual-coolant blanket within the framework of the EU power plant conceptual study (TW2-TRP-PPCS12), Final Report, Forschungszentrum Karlsruhe, Wissenschaftliche Berichte, FZKA 6780, 2003.

[2.1-6] L. Giancarli, L. Bühler, U. Fischer, R. Enderle, D. Maisonnier, C. Pascal, et al., Invessel component designs for a self-cooled lithium-/lead fusion reactor, Fusion Engineering and Design 69 (2003) 763–768.

[2.1-7] A.R. Raffray, L. El-Guebaly, T. Ihli, S. Malang, F. Najmabadi, X.Wang, Engineering challenges in designing an attractive compact stellarator power plant, Fusion Engineering and Design 82 (2007) 2696–2704.

[2.2-1] R. Tivey, M. Akiba, D. Driemeyer, I. Mazul, M. Merola, M. Ulrickson, ITER R&D: Vacuum Vessel and In-Vessel Components: Divertor Cassette, Fusion Engineering and Design 55 (2001) 219–229.

[2.2-2] L.V. Boccaccini, S. Hermsmeyer, U. Fischer, T. Ihli, G. Janeschitz, C. Köhly, D. Nagy, P. Norajitra, C. Polixa, J. Reimann, J. Rey, Studies of In-Vessel Component Integration for a Helium-cooled Fusion Reactor, Proceedings of the IAEA-TM1, IAEA, Vienna, 2006, ISBN 92-0-106306-7.

[2.2-3] H. Bolt, A. Brendel, D. Levchuk, H. Greuner and H. Maier, Materials for the plasma-facing components of fusion reactors, Energy Materials, Vol. 1 No. 2 (2006) 121-126.

[2.2-4] ITER Nuclear Analysis Report, G 73 DDD 2 01-06-06 W0.1, June 2001.

[2.2-5] M. Ferrari, L. Giancarli, K. Kleefeldt, C. Nardi, M. Rödig, J. Reimann, J.F. Salavy, Evaluation of divertor conceptual designs for a fusion power plant, Fusion Engineering and Design 56–57 (2001) 255–259.

[2.2-6] V. Barabash, A. Peacock, S. Fabritsiev, G. Kalinin, S. Zinkle, A. Rowcliffe, et al., Materials challenges for ITER – Current status and future activities, Journal of Nuclear Materials 367–370 (2007) 21–32.

[2.3-1] A. Li Puma, L. Giancarli, Y. Poitevin, J.-F. Salavy, P. Sardain, J. Szczepanski, Optimization of a water-cooled divertor for the European power plant conceptual study, Fusion Engineering and Design 61–62 (2002) 177–183.

[2.3-2] L. Giancarli, J.P. Bonal, A. Li Puma, B. Michel, P. Sardain, J.F. Salavy, Conceptual design of a high temperature water-cooled divertor for a fusion power reactor, Fusion Engineering and Design 75–79 (2005) 383–386.

[2.3-3] A. Li Puma, L. Giancarli, H. Golfier, Y. Poitevin, J. Szczepanski, Potential performances of a divertor concept based on liquid metal cooled SiC_f/SiC structures, Fusion Engineering and Design 66–68 (2003) 401–405.

[2.3-4] K. Kleefeldt, S. Gordeev, Performance limits of a helium-cooled divertor (unconventional design), Forschungszentrum Karlsruhe, Wissenschaftliche Berichte, FZKA 6401, 2000.

[2.3-5] S. Hermsmeyer, K. Kleefeldt, Review and comparative assessment of helium cooled divertor concepts, Forschungszentrum Karlsruhe, Wissenschaftliche Berichte, FZKA 6597, 2001.

[2.3-6] S. Hermsmeyer, S. Malang, Gas-cooled high performance divertor for a power plant, Fusion Engineering and Design 61–62 (2002) 197–202.

[3.1-1] G. Janeschitz et al., Divertor development for ITER, Fusion Engineering and Design 39-40 (1998) 173-187.

[3.1-2] G. Janeschitz et al., Overview of the divertor design and its integration into RTO:RC-ITER G, Fusion Engineering and Design 49–50 (2000) 107–117.

[3.2-1]   http://yarchive.net/nuke/nuclear_plant_startup.html.   Retrieved   March   31, 2010.

[3.4-1] H. Meister, R Fischer, L.D. Horton, C.F. Maggi, D. Nishijima, C. Giroud, K.-D. Zastrow, B. Zaniol, et al., Zeff from spectroscopic bremsstrahlung measurements at ASDEX Upgrade and JET, 33rd EPS Conference on Plasma Phys. Rome, 19 - 23 June 2006 ECA Vol.30I, P-2.138 (2006).

[3.4-2] R. Behrisch, W. Eckstein, Sputtering by Particle Bombardment Experiments, Experiments and Computer Calculations from Threshold to MeV Energies, Springer, Berlin, 2007, ISBN: 978-3-540-44500-5.

[3.4-3] W. Eckstein, C. Garcia-Rosales, J. Roth and W. Ottenberger, Sputtering data, IPP Tech. Rep. IPP 8/82 Max-Planck-Institut für Plasmaphysik, Garching (1993).

[3.4-4] U. Fischer, P. Pereslavtsev, A. Möslang, M. Rieth, Transmutation and activation analysis for divertor materials in a HCLL-type fusion power reactor, Journal of Nuclear Materials 386–388 (2009) 789–792.

[3.4-5] V. Massaut, R. Bestwick, K. Broden, L. Di Pace, L. Ooms, R. Pampin, State of the art of fusion material recycling and remaining issues, Fusion Engineering and Design 82 (2007) 2844–2849.

[3.6-1] P. Norajitra, R. Giniyatulin, G. Janeschitz, D. Maisonnier, I. Mazul, A. Pizzuto, et al., European development of He-cooled divertors for fusion power plants, Nuclear Fusion 45 (2005) 1271–1276.

[3.6-2] T. Ihli, R. Kruessmann, I. Ovchinnikov, P. Norajitra, V. Kuznetsov, R. Giniyatulin, An advanced He-cooled divertor concept: design, cooling technology, and thermohydraulic analyses with CFD, Fusion Engineering and Design 75–79 (2005) 371–375.

[3.6-3] P. Norajitra, R. Giniyatulin, T. Ihli, G. Janeschitz, V. Kuznetsov, I. Mazul, et al., He-cooled divertor development for DEMO, Fusion Engineering and Design 82 (2007) 2740–2744.

[3.7-1] R. Kruessmann, P. Norajitra, L.V. Boccaccini, T. Chehtov, R. Giniyatulin, S. Gordeev, T. Ihli, G. Janeschitz, A.O. Komarov, W. Krauss, V. Kuznetsov, R. Lindau, I. Ovchinnikov, V. Piotter, M. Rieth, R. Ruprecht, V. Slobodtchouk, V.A. Smirnov, R. Sunyk, Conceptual design of a He-cooled divertor with integrated flow and heat transfer promoters. (PPCS subtask TW3-TRP-001-D2) Part I: Summary Wissenschaftliche Berichte, FZKA-6974 (April 2004).

[3.7-2] R. Kruessmann, P. Norajitra, L.V. Boccaccini, T. Chehtov, R. Giniyatulin, S. Gordeev, T. Ihli, G. Janeschitz, A.O. Komarov, W. Krauss, V. Kuznetsov, R. Lindau, I. Ovchinnikov, V. Piotter, M. Rieth, R. Ruprecht, V. Slobodtchouk, V.A. Smirnov, R. Sunyk, Conceptual design of a He-cooled divertor with integrated flow and heat transfer promoters. (PPCS subtask TW3-TRP-001-D2) Part II: Detailed Version. Wissenschaftliche Berichte, FZKA-6975 (April 2004).

[3.7-3] VDI-Wärmeatlas, Zehnte, bearbeitete und erweiterte Auflage, ©Springer-Verlag, Berlin Heidelberg 2006.

[3.7-4] M. Dalle Donne, S. Dorner, S. Taczanowski, Conceptual design of two Helium cooled fusion blankets (ceramic and liquid breeder) for INTOR, Kernforschungszentrum Karlsruhe; KfK-3584 (EUR-7987e), 1983.

[3.7-5] P. Norajitra, T. Chehtov, A. Gervash, R. Giniyatulin, T. Ihli, R. Kruessmann, V. Kuznetsov, A. Makhankov, I. Mazul, I. Ovchinnikov, J. Weggen, B. Zeep, Status of He-cooled divertor development (PPCS subtask TW4-TRP-001-D2). Wissenschaftliche Berichte, FZKA-7100 (Februar 2005).

[3.7-6] S. Gordeev, F. Arbeiter, V. Heinzel, V. Slobodtchouk, Analysis of turbulence models for thermohydraulic calculations of helium cooled fusion reactor components, Fusion Engineering and Design 81 (2006) 1555–1560.

[3.8-1] R. Krüssmann, T. Ihli, P. Norajitra, "Divertor cooling concept: CFD parametric study for design optimization", Jahrestagung der Kerntechnischen Gesellschaft Deutschland, Aachen, May 2006, Proceedings: INFORUM GmbH, Berlin (2006) 558 - 561.

[3.8-2] ANSYS CFX 10 and 11, ANSYS Inc., 2006.

[3.8-3] I. Ovchinnikov, R. Giniyatulin, A. Komarov, V. Kuznetsov, S. Mikhailov, V. Smirnov, et al., Experimental study of DEMO helium cooled divertor target mock-ups to estimate their thermal and pumping efficiencies, Fusion Engineering and Design 73 (2005) 181–186.

[3.8-4] L. Crosatti, J. B. Weathers, D. L. Sadowski, S. I. Abdel-Khalik, M. Yoda, R. Kruessmann, P. Norajitra, Experimental and Numerical Investigation of Prototypical Multi-Jet Impingement (HEMJ) Helium-Cooled Divertor Modules, Fusion Science and Technology, Volume 56, Number 1, July 2009, Pages 70-74.

[3.8-5] R. Krüßmann, G. Messemer, K. Zinn, Overview of thermohydraulic simulations for the development of a helium-cooled divertor, Forschungszentrum Karlsruhe, Wissenschaftliche Berichte, FZKA 7394, 2008.

[3.8-6] ANSYS Inc., ANSYS Release 12.0 Documentation, 2009.

[3.8-7] V. Widak, P. Norajitra, Optimization of He-cooled divertor cooling fingers using a CAD-FEM method, Fusion Engineering and Design 84 (2009) 1973–1978.

[3.8-8] ASME Code III, edition 1986.

[3.8-9] P. Norajitra, Thermohydraulics Design and Thermomechanics Analysis of Two European Breeder Blanket Concepts for DEMO, Forschungszentrum Karlsruhe, Wissenschaftliche Berichte, FZK 5580, 1995.

[3.8-10] ITER Materials Properties Handbook (MPH), ITER Doc. G 74 MA 16 04-05-07 R0.1.

[3.9-1] R. Giniyatulin, A. Gervash, W. Krauss, A. Makhankov, I. Mazul, P. Norajitra, Study of technological and material aspects of He-cooled divertor for DEMO reactor, 23rd SOFT, Venice, Italy, 20.–24.9.2004.

[3.9-2] P. Norajitra, R. Giniyatulin, T. Ihli, G. Janeschitz, W. Krauss, R. Kruessmann, V. Kuznetsov, I. Mazul, V. Widak, I. Ovchinnikov, R. Ruprecht, B. Zeep, He-cooled divertor development for DEMO, Fusion Eng. Design 82 (2007) 2740–2744.

[3.9-3] P. Norajitra, A. Gervash, R. Giniyatulin, T. Hirai, G. Janeschitz, W. Krauss, V. Kuznetsov, A. Makhankov, I. Mazul, I. Ovchinnikov, J. Reiser, V. Widak, Helium-cooled divertor for DEMO: Manufacture and high heat flux tests of tungsten-based mock-ups, Journal of Nuclear Materials 386–388 (2009) 813–816.

[3.9-4] P. Norajitra, R. Giniyatulin, T. Hirai, W. Krauss, V. Kuznetsov, I. Mazul, I. Ovchinnikov, J. Reiser, G. Ritz, H.-J. Ritzhaupt-Kleissl, V.Widak, Current status of He-cooled divertor development for DEMO, Fusion Engineering and Design 84 (2009) 1429–1433.

[3.9-5] G. Ritz, T. Hirai, J. Linke, P. Norajitra, R. Giniyatulin, Post-examination of helium-cooled tungsten modules exposed to cyclic thermal loads, in: Jahrestagung der Kerntechnischen Gesellschaft Deutschland, Hamburg, May 2008, Proceedings: INFORUM GmbH, Berlin, 2008, pp. 685–688.

[4-1] M. Rieth et al., Review on the EFDA Programme on Divertor Materials Technology and Science, Journal of Nuclear Materials (2011), doi: 10.1016/j.jnucmat.2011.01.075.

[4.1-1] G. Ritz, T. Hirai, J. Linke, P. Norajitra, R. Giniyatulin, L. Singheiser, Postexamination of helium-cooled tungsten components exposed to DEMO specificcyclic thermal loads, Fusion Engineering and Design 84 (2009) 1623–1627.

[4.1-2] P. Norajitra, J. Reiser, H.-J. Ritzhaupt-Kleissl, S. Dichiser, J. Konrad, G. Ritz, Development of a He-Cooled Divertor: Status of the Fabrication Technology, Fusion Science and Technology, Volume 56, Number 1, July 2009, Pages 80-84.

[4.2-1] P. Norajitra, S. Antusch, R. Giniyatulin, I. Mazul, G. Ritz, H.-J. Ritzhaupt-Kleissl, L. Spatafora, Current State-Of-The-Art Manufacturing Technology for He-Cooled Divertor Finger, J. Nucl. Mater. 417 (2011) 468–471.

[4.2-2] P. Norajitra, S. Antusch, R. Giniyatulin, V. Kuznetsov, I. Mazul, H.-J. Ritzhaupt-Kleissl, L. Spatafora, Progress of He-cooled Divertor Development for DEMO, Fusion Eng. Des. 86 (2011) 1656–1659.

[4.3-1] S. Antusch, P. Norajitra, V. Piotter, H.-J. Ritzhaupt-Kleissl, L. Spatafora, Powder Injection Molding – an innovative manufacturing method for He-cooled DEMO Divertor components, Fusion Eng. Des. 86 (2011) 1575–1578.

C Nomenclature

A_i (m^2)	Cross-sectional area of the jet hole
B [m]	Jet slot width
B [T] [Vs/m^2]	Magnetic field
c [m/s]	speed of light (299792458 m/s)
c_p [J/kgK]	Specific heat capacity
c_{Br}	Bremsstrahlung constant
c_{He} [m/s]	Speed of sound in helium
D [m]	Jet diameter, hydraulic diameter
E [J]	Energy
E [MPa]	Young's modulus (ch. 5.3.3)
G	Geometry function
K [J/K]	Boltzmann constant = $1.38.10^{-23}$ J/K
L_T [m]	Pitch distance
m [kg]	Mass
m_i [kg/s]	Jet mass flow rate
n [m^{-3}]	Plasma particle density (nuclei per plasma unit volume)
n_e, n_i [m^{-3}]	Number of electrons or ions per plasma unit volume ($n_e = n_i = n$) (ch. 2.5.2)
n_D [m^{-3}]	Deuterons density
n_T [m^{-3}]	Tritons density
p [MPa]	Pressure, plasma pressure
Δp [MPa]	Pressure loss
P [W]	Power
Pr [-]	Prandtl number
P_{Br} [W]	Radiation power by bremsstrahlung per unit volume of plasma
q	Safety factor
q [W/m^2]	Heat flux
q [As]	char ge of the carrier = $\pm 1.602 \times 10^{-19}$ As
Q [-]	Power amplification factor or energy gain (i.e. the ratio between the power from fusion reactions and the external power supplied to the plasma by the heating systems)
$Q_{aux.\ heating}$ [W]	Auxiliary heating power
$Q_{neutron}$ [W]	Neutron volumetric heat power
Q_{surf} [W]	Surface heat power (= $Q_\alpha + Q_{aux.\ heating}$)
Q_α [W]	Alpha article power
r_g [m]	radius of gyration
R [J/kgK]	Gas constant = 2078.75 J/kgK for helium
Re [-]	Reynolds number
R_{12} [1/s]	Reaction rate
Rm [MPa]	Ultimate tensile strength

$R_{p0.2}$ [MPa]	Offset yield strength
$R_{p1.0,t}$ [MPa]	1 % proof stress in time t
$R_{u,t}$ [MPa]	creep rupture strength in time t
S_m, $S_{m,t}$ [MPa]	Value to evaluate the stress results based on the maximum stress theory used by ASME code
T [K] [°C]	Temperature
T_e [K]	Electron temperature
α, TEC [1/K]	Thermal linear expansion coefficient
v [m/s]	Velocity, relative velocity
V_C [MeV]	Height of the Coulomb barrier (ch. 2-3)
w [m/s]	Average jet velocity
Z_i	Atomic number of plasma ions ($Z_i = 1$ for hydrogen plasmas)
β	Ratio of the plasma kinetic pressure (proportional to its density and temperature) to the confinement magnetic pressure (proportional to the intensity of the magnetic field)
η [kg/ms]	Dynamic viscosity
κ	Adiabatic exponent, isentropic exponent, k-value
λ [W/mK]	Thermal conductivity
v [m²/s]	Kinematic viscosity
v [-]	Poisson's ratio
ρ [kg/m3]	Density
ρ_{el} [Ω.m]	Electrical resistivity, specific electrical resistance
σ_{eq} [MPa]	Equivalent stress intensity, here: von Mises stress (ch. 5.5.3)
σ_F [m²]	Fusion cross-section [1 m² = 10^{28} barn]
σ_T [MPa.m/W]	Stress factor
τ_E [s]	Energy confinement time

Abbreviations

A	Mass number (A = N+Z)
AMC	Active metal cast
ARIES-CS	Advanced Reactor Innovation and Evaluation Study-Compact stellarator
ARIES-ST	Advanced Reactor Innovation and Evaluation Study-Spherical Torus
ASDEX	Axially Symmetric Divertor Experiment
ASME	American Society of Mechanical Engineers
Av.	Average
A	Mass number (A = N+Z)
B	Boron
Be	Beryllium
Ba	Barium
BMWi	Bundesministerium für Wirtschaft und Technologie
CuCrZr	Copper alloy (copper chromium zirconium)
CuNi44	Copper Nickel brazing alloy material
CAD	Computer Aided Design
CFC	Carbon fiber composite
CFD	Computational Fluid Dynamics
DIII-D	The fusion experiment Doublet
DBTT	Ductile-brittle transition temperature
DEMO	Demonstration reactor
D	Deuterium
dpa	Displacements per atom
e	Electron
EB	Electron beam
EDM	Electro discharge machining
EFDA	European Fusion Development Agreement
ELMs	Edge-Localised-Modes
ELM coils	Magnetic coils that provide a magnetic "massage" of the plasma exterior to suppress potentially harmful power deposition on plasma-facing components
ENEA	Italian National Agency for New Technologies, Energy and Sustainable Economic Development
EU	European Union
EUROFER	Reduced-activation ferritic steel
Fe	Iron
FEM	Finite Element Method
FPP	Fusion Power Plant
FPY	Full-power year
FW	First wall
GT	G.W. Woodruff School of Mechanical Engineering, Georgia

	Institute of Technology, Atlanta, GA 0332-0405, USA
H	Hydrogen
HCD	Helium cooled divertor
He	Helium
HEBLO	Helium Blanket Test Loop
Hf	Hafnium
KIT	Karlsruhe Institute of Technology
HCLL	Helium-cooled liquid lead-lithium (blanket concept)
HCPB	Helium-cooled pebble bed (blanket concept)
He	Helium
HEMJ	He-cooled modular divertor with jet cooling
HEMP	Helium-cooled modular divertor concept with integrated pin array
HEMS	He-cooled modular divertor with slot array
HETS	High-efficiency thermal shield
HEX	Recuperator, heat exchanger
HHF	High heat flux
HIP	Hot isostatic pressing
HT	High-temperature
htc	Heat transfer coefficient
H-mode	High confinement mode
IB	Inboard
Ir	Iridium
IR	Infrared
ITER	International Thermonuclear Experimental Reactor
JET	Joint European Torus
JT-60	JAERI (Japan Atomic Energy Research Institute) Tokamak
Kr	Krypton
Li	Lithium
La	Lanthanium
La_2O_3	Lanthanum Oxide
LCFS	Last closed flux surface
L-mode	Low confinement mode
LMCD	Liquid metal cooled divertor
LT	Low-temperature
mfr, MFR	Mass flow rate
MARFE	Multifaceted Asymmetric Radiation From the Edge
Max.	Maximum
MHD	Magneto-hydrodynamic
MPH	Material Properties Handbook
Mo	Molybdenum
n	Neutron
N	Neutron number of a nucleus ($N = A - Z$)
Nb	Niobium
Nb3Sn	Triniobium-tin (type II superconductor)

NbTi	Niobium-titanium (type II superconductor)
Nu	Nusselt number
OB	Outboard
ODS	Oxide dispersion-strengthened
OFHC	Oxygen free high conductivity
Pb-17Li	Eutectic lead-lithium alloy
p	Proton
P	Private flux region
PdCu	Palladium-copper brazing filler
PdNi	Palladium Nickel brazing filler
PFM	Plasma facing material
PHF	Peak heat flux
pol	Poloidal
PPCS	Power plant conceptual study
PV	Photovoltaic
PWR	(Fission) pressurized water reactor
rad	Radial
RAFM	Reduced-activation ferritic/martensitic
R&D	Research and Development
RCT	Recrystallization temperature
Re	Rhenium
Rh	Rhodium
RMP	Resonant Magnetic Perturbation
RNG	Reynolds normalisation group
S	Separatrix
SCLL	Self-cooled liquid-lead (blanket concept)
SiC_f/SiC	Silicon carbide composite
SKE	Coal equivalent
SOL	Scrape-off layer
SPD	Severe plastic deformation
STEMET® 1311	Ni-based brazing filler material
tor	Toroidal
T	Tritium
TEC	Thermal expansion coefficient
Ta	Tantalium
Ti	Titanium
T91	Ferritic steel
TBM	Test blanket module
TC	Thermocouple
TCV	Tokamak à Configuration Variable
TDM	Test divertor module
TSEFEY	Electron beam facility
ThO_2	Thorium dioxide

TZM	Molybdenum alloy with 0.5 % Ti, 0.08 % Zr, and 0.04 % C
U	Uranium
US	United States
VDE	Vertical Displacement Event
VS coils	Magnetic coils that provide fast vertical stabilization of the plasma
VT	Vertical target
W	Tungsten
WCD	Water cooled divertor
WCLL	Water-cooled liquid lead-lithium
WL10	W 1.0 % (by weight) La_2O_3
Y_2O_3	Yttrium oxide
Z	Atomic number or proton number
Z_{eff}	Effectiveatomicnumber (ch. 3.3). The effective ionic charge Zeff is a means to assess the impurity content of a fusion plasma.
71KHCP®	Co-based brazing filler

Subscript

br	brazing
in	inlet
liq	liquidus
out	outlet

Acknowledgements

This work, supported by the European Communities under the contract of association between EURATOM and Karlsruhe Institute of Technology (KIT), was carried out during my activity as a researcher and as divertor project leader at the Institute for Applied Materials - Materials Processing Technology (IAM-WPT) at KIT within the framework of the European Fusion Development Agreement. The views and opinions expressed herein do not necessarily reflect those of the European Commission.

The project to develop a divertor of a future fusion power plant is a challenging and highly complex task that could not have run without valuable professional support and cooperation of numerous colleagues at home and abroad. I would therefore like to thank:

Prof. Dr. Oliver Kraft and Prof. Dr. Jürgen Haußelt for making possible the thesis at the Institute for Applied Materials - Materials and Biomechanics (IAM-WBM) and Materials Processing Technology (IAM-WPT) at KIT and for reviewing my manuscript and numerous informative notes.

Dr. Günter Janeschitz for his technical scientific support and valuable discussions.

Dipl.-Ing. Siegfried Malang for many fruitful discussions and valuable advices.

Professor Dr. Robert Stieglitz and Dr. Hans-Joachim Ritzhaupt-Kleissl for reviewing my manuscript and numerous hints. Prof. Dr. Thomas Hanemann and Dr. Robert Ruprecht for many valuable suggestions.

Dr. Thomas Ihli, Dr. Regina Krüßmann, Dipl.-Ing. Jens Reiser, B.Eng. Luigi Spatafora, and Dipl.-Ing. (FH) Verena Widak for their fruitful cooperation in the divertor team.

The Efremov Team, St. Petersburg, Russia, especially Dr. Alexander Gervash, Dr. Radmir Giniyatulin, Dr. Vladimir Kuznetsov, Dr. Alexey Makhankov, Dr. Igor Mazul, and Dr. Ivan Ovchnnikov for their good and fruitful cooperation and valuable high-heat-flux experiments.

Dr. Jarir Aktaa, Dr. Steffen Antusch, Dr. Widodo Basuki, Mr. Steffen Berberich, Dr. Lorenzo Boccaccini, Dipl.-Ing. Henri Greuner (IPP/Garching), Dr. Takeshi Hirai (ITER), Dr. Wolfgang Krauss, Dr. Jochen Linke (FZJ), Dr. Anton Möslang, Dr. Rudolf Neu (IPP/Garching), and Dr. Michael Rieth for many valuable discussions.

Mr. Stefan Dichiser and Mr. Joachim Konrad for their good and agreeable cooperation in the study of manufacturing and joining of the divertor components.

Mr. Bernhard Dafferner, Mr. Momir Kuzmic, and Mr. Volker Toth for all sorts of technical support in the design accompanying tests.

Karlsdorf-Neuthard,   *Prachai Norajitra*

in September 2011